PENGUIN BOOKS

SYNC

Steven Strogatz is Professor of Applied Mathematics at Cornell University and one of the world's leading researchers into chaos, complexity and synchronization. His seminal research has been featured in *Nature*, *Science*, *Scientific American*, *The New York Times*, the *New Yorker*, and the *Daily Telegraph*.

STEVEN STROGATZ

SYNC

The Emerging Science of
Spontaneous Order

PENGUIN BOOKS

PENGUIN BOOKS

Published by the Penguin Group
Penguin Books Ltd, 80 Strand, London WC2R 0RL, England
Penguin Group (USA), Inc., 375 Hudson Street, New York, New York 10014, USA
Penguin Books Australia Ltd, 250 Camberwell Road, Camberwell, Victoria 3124, Australia
Penguin Books Canada Ltd, 10 Alcorn Avenue, Toronto, Ontario, Canada M4V 3B2
Penguin Books India (P) Ltd, 11 Community Centre, Panchsheel Park, New Delhi – 110 017, India
Penguin Books (NZ) Ltd, Cnr Rosedale and Airborne Roads, Albany, Auckland, New Zealand
Penguin Books (South Africa) (Pty) Ltd, 24 Sturdee Avenue, Rosebank 2196, South Africa

Penguin Books Ltd, Registered Offices: 80 Strand, London WC2R 0RL, England

www.penguin.com

Published in the United States of America by Hyperion 2003
Published in Great Britain by Allen Lane 2003
Published in Penguin Books 2004
1

Grateful acknowledgement is made for permission to reprint excerpts from the following copyrighted works:
'Out of Step on the Bridge' and 'Physics and the Nobel Prizes' copyright © Brian Josephson, 2000, 2001.
Reprinted by permission of Brian Josephson

Illustrations by Margaret Nelson

Illustration credits p.80 – Figure 1.9 of Martin Moore-Ede, Frank M. Sulzman, and Charles A. Fuller,
The Clocks That Time Us (Cambridge, Massachusetts: Harvard University Press, 1982), adapted by
permission of Martin Moore Ede; p. 282 – Figure 1 of Eugenio Rodriguez et al., 'Perception's Shadow:
Long-distance synchronization of human brain activity', *Nature* 397 (1999) pp. 430–33, adapted
by permission of Jacques Martinerie

Printed in England by Clays Ltd, w Ives plc

To Art Winfree
Mentor, inspiration, friend

CONTENTS

III · EXPLORING SYNC

SYNC

PREFACE

AT THE HEART OF THE UNIVERSE IS a steady, insistent beat: the sound of cycles in sync. It pervades nature at every scale from the nucleus to the cosmos. Every night along the tidal rivers of Malaysia, thousands of fireflies congregate in the mangroves and flash in unison, without any leader or cue from the environment. Trillions of electrons march in lockstep in a superconductor, enabling electricity to flow through it with zero resistance. In the solar system, gravitational synchrony can eject huge boulders out of the asteroid belt and toward Earth; the cataclysmic impact of one such meteor is thought to have killed the dinosaurs. Even our bodies are symphonies of rhythm, kept alive by the relentless, coordinated firing of thousands of pacemaker cells in our hearts. In every case, these feats of synchrony occur spontaneously, almost as if nature has an eerie yearning for order.

And that raises a profound mystery: Scientists have long been baffled by the existence of spontaneous order in the universe. The laws of thermodynamics seem to dictate the opposite, that nature should inexorably degenerate toward a state of greater disorder, greater entropy. Yet all around us we see magnificent structures—galaxies, cells, ecosystems, human beings—that have somehow managed to assemble themselves. This enigma bedevils all of science today.

Only in a few situations do we have a clear understanding of how order arises on its own. The first case to yield was a particular kind of order *in physical space* involving perfectly repetitive architectures. It's the kind of order that occurs whenever the temperature drops below the freezing point and trillions of water molecules spontaneously lock themselves into a rigid, symmetrical crystal of ice. Explaining order *in time*, however, has proved to be more problematic. Even the simplest possibility, where the same things happen at the same times, has turned out to be remarkably subtle. This is the order we call synchrony.

It may seem at first that there's little to explain. You can agree to meet a friend at a restaurant, and if both of you are punctual, your arrivals will be synchronized. An equally mundane kind of synchrony is triggered by a reaction to a common stimulus. Pigeons startled by a car backfiring will all take off at the same time, and their wings may even flap in sync for a while, but only because they reacted the same way to the same noise. They're not actually communicating about their flapping rhythm and don't maintain their synchrony after the first few seconds. Other kinds of transient sync can arise by chance. On a Sunday morning, the bells of two different churches may happen to ring at the same time for a while, and then drift apart. Or while sitting in your car, waiting to turn at a red light, you might notice that your blinker is flashing in perfect time with that of the car ahead of you, at least for a few beats. Such sync is pure coincidence, and hardly worth noting.

The impressive kind of sync is persistent. When two things keep happening simultaneously for an extended period of time, the synchrony is probably not an accident. Such persistent sync comes easily to us human beings, and, for some reason, it often gives us pleasure. We like to dance together, sing in a choir, play in a band. In its most refined form, persistent sync can be spectacular, as in the kickline of the Rockettes or the matched movements of synchronized swimmers. The feeling of artistry is heightened when the audience has no idea where the music is going next, or what the next dance move will be. We interpret persistent sync as a sign of intelligence, planning, and choreography.

So when sync occurs among unconscious entities like electrons or cells, it seems almost miraculous. It's surprising enough to see animals cooperating—

thousands of crickets chirping in unison on a summer night; the graceful undulating of schools of fish—but it's even more shocking to see mobs of mindless things falling into step by themselves. These phenomena are so incredible that some commentators have been led to deny their existence, attributing them to illusions, accidents, or perceptual errors. Other observers have soared into mysticism, attributing sync to supernatural forces in the cosmos.

Until just a few years ago, the study of synchrony was a splintered affair, with biologists, physicists, mathematicians, astronomers, engineers, and sociologists laboring in their separate fields, pursuing seemingly independent lines of inquiry. Yet little by little, a science of sync has begun coalescing out of insights from these and other disciplines. This new science centers on the study of "coupled oscillators." Groups of fireflies, planets, or pacemaker cells are all collections of oscillators—entities that cycle automatically, that repeat themselves over and over again at more or less regular time intervals. Fireflies flash; planets orbit; pacemaker cells fire. Two or more oscillators are said to be coupled if some physical or chemical process allows them to influence one another. Fireflies communicate with light. Planets tug on one another with gravity. Heart cells pass electrical currents back and forth. As these examples suggest, nature uses every available channel to allow its oscillators to talk to one another. And the result of those conversations is often synchrony, in which all the oscillators begin to move as one.

Those of us working in this emerging field are asking such questions as: How exactly do coupled oscillators synchronize themselves, and under what conditions? When is sync impossible and when is it inevitable? What other modes of organization are to be expected when sync breaks down? And what are the practical implications of all that we're trying to learn?

I've been fascinated by such questions for 20 years, first as a graduate student at Harvard University and then as a professor of applied math at the Massachusetts Institute of Technology and Cornell University, where I now teach and do research on chaos and complexity theory. My interest in cycles goes back even further than that, to an epiphany I had as a freshman in high school. For one of the first experiments in Science I, Mr. diCurcio gave each of us a stopwatch and a little toy pendulum, a tricky gadget with an extensible arm that could be lengthened or shortened in discrete steps, like one of those old telescopes you see in pirate movies. Our assignment was to clock the pendulum's period—the time it takes for one swing back and forth—and to figure out how its period depends on its length: Does a longer pendulum swing faster, slower, or stay the same? To find out, we set our pendulums to the shortest length, timed its period, and plotted the result on a piece of graph paper. Then we repeated the experiment for progressively longer pendulums, always stretching the arm one click at a time. As I drew the fourth or fifth dot on the graph paper, it suddenly dawned on me that a pattern was emerging: The dots were falling on a parabolic curve. The same parabolas that I was learning about in Algebra II were secretly governing the motions of these pendulums. An enveloping sensation of wonder and fear came over me. In that moment of revelation, I became aware of a hidden but beautiful world that can be seen only through mathematics. It was a moment from which I have never really recovered.

Thirty years later, I'm still captivated by the mathematics of nature, especially as manifested by things that move in cycles, like the periodic swaying of the pendulum. But instead of a single cycle, my research has taken me to the study of many of them working together all at once—to the study of coupled oscillators. My training leads me to make simple models, to replace the bewildering complexity and richness of real fireflies or superconductors with idealized sets of equations that mimic their group behavior. I try to use calculus and computers to see how order emerges from chaos. What makes these puzzles so much fun is that they lie at the edge of known mathematics. Two coupled oscillators would be no challenge—their behavior has been understood since the early 1950s. But for questions involving hundreds or thousands of oscillators, we're still in the dark. The nonlinear dynamics of systems with that many variables is still beyond us. Even with the help of supercomputers, the collective behavior of gigantic systems of oscillators remains a forbidding terra incognita.

Still, over the past decade, thanks to the combined efforts of mathematicians and physicists around the world, one special case has finally been worked out, opening the door to a deeper understanding of sync. If we assume that all the oscillators in a given group are nearly identical, and that they are all coupled equally to one another, the dynamics become mathematically tractable. In Parts I and II of this book, I tell the story of how my colleagues and I solved this class of theoretical problems, and what their solutions imply for sync in the real world: in Part I for living oscillators (cells, animals, and people) and then in Part II with reference to inanimate oscillators (pendulums, planets, lasers, and electrons). Part III deals with the frontiers of sync, when we cast aside our earlier simplifying assumptions. This realm is still largely unexplored, and includes situations where the oscillators are replaced by chaotic systems, or where they are coupled in less symmetrical ways—to their neighbors in three-dimensional space, or in intricate networks that transcend geography.

Sync is an attempt to synthesize a vast body of knowledge on this subject created by scientists working across disciplines, continents, and centuries. The science needed to understand sync draws on the work of some of the greatest minds of the twentieth century, many of whom are household names and others who should be—the physicists Albert Einstein, Richard Feynman, Brian Josephson, and Yoshiki Kuramoto; the mathematicians Norbert Wiener and Paul Erdős; the social psychologist Stanley Milgram; the chemist Boris Belousov; the chaos theorist Edward Lorenz; and the biologists Charles Czeisler and Arthur Winfree.

My own research runs through the story, not because I have any illusions about my place in history, but because I want to give a feel for what it's like to be working in the trenches of science—the blind alleys, the twists and turns, the exhilaration of discovery, the metamorphosis from student to colleague to mentor. To convey the vitality of mathematics to a broad spectrum of readers, I've avoided equations altogether, and rely instead on metaphors and images from everyday life to illustrate the key ideas.

My hope is that you'll come to share some of my excitement about the breathtaking diversity of synchronization in the natural world, and the power of mathematics to explain it. Sync is both strange and beautiful. It is strange because it seems to defy the laws of physics (though in fact it relies on them, often in curious ways). It is beautiful because it results in a kind of cosmic ballet that plays out on stages that range from our bodies to the universe as a whole. And it is also critically important. Our basic understanding of sync has already spawned such technological wonders as the global positioning system; the laser; and the world's most sensitive detectors, used by doctors to pinpoint diseased tissues in the brains of epileptics without the need for surgery, by engineers to search for tiny cracks in airplane wings, and by geologists to locate oil buried deep underground. By investigating what happens when sync unravels, mathematicians are helping cardiologists track down the cause of fibrillation, a deadly arrhythmia that kills hundreds of thousands of people every year, suddenly and without warning, even those with no history of heart disease. And this is just a sample of what we are able to do today, thanks to our growing but still rudimentary knowledge of sync.

I am deeply grateful for the opportunity to have worked with so many brilliant and creative minds throughout my career. The research described here was a joint effort with my advisers Art Winfree, Richard Kronauer, Chuck Czeisler, and Nancy Kopell; my collaborators Rennie Mirollo, Paul Matthews, Kurt Wiesenfeld, Jim Swift, Kevin Cuomo, Al Oppenheim, and Tim Forrest; and my former students Shinya Watanabe and Duncan Watts. Thanks for being such wonderful companions on our journeys into the wilds of sync.

Other scientists helped improve the book in various ways. Jack Cowan shared his affectionate memories of Norbert Wiener at MIT in the late 1950s and enlightened me with the untold but very human story behind the double-dip spectrum. Lou Pecora provided a blow-by-blow account of how he and Tom Carroll were led to the discovery of synchronized chaos. Jim Thorp answered my questions about the power grid with his usual wisdom and good humor. Cedric Langbort kindly translated Huygens's correspondence about the sympathy of clocks. Joe Burns, Erik Herzog, Chris Lobb, Charlie Marcus, Raj Roy, and Joe Takahashi offered insightful comments on early drafts of the manuscript. Margy Nelson prepared the illustrations with her distinctive blend of scientific judgment and artistic flair. I'm especially grateful to Art Winfree for sharing his playfulness and his mastery of sync, and, above all, for his heroic and amazingly generous effort in reading the manuscript from cover to cover, even under the most difficult circumstances.

Thank you to Lindy Williams, Stephen Tien, Herbert Hui, Tom Gilovich, and all my other friends who so patiently endured my tribulations in the early stages; Karen Dashiff Gilovich, who helped me find my voice; and Alan Alda, a terrifically stimulating partner in brainstorming sessions, who taught me a lot about how to approach the creative process. (Though I never did manage to follow his best piece of advice, about writing the first draft in one long, happy belch. Maybe next time.)

My colleagues at Cornell, especially Richard Rand and my department chairman, Tim Healey, have provided encouragement and support throughout the exhausting process of writing this book and have been patient with me whenever my mind seemed to be elsewhere. Thanks for being so understanding.

My literary agents Katinka Matson and John Brockman have been enthusiastic and helpful at every turn. John suggested the main title for the book within a millisecond of hearing my description of it. Katinka gently coached me through all aspects of the book-writing process, from proposal to publication.

A writer could not ask for a better publication team than the staff at Hyperion Books. In particular, editorial assistant Kiera Hepford was always gracious, upbeat, and efficient. Art director Phil Rose designed a cover that captures the essence of sync memorably and beautifully. And thanks especially to my editor, Will Schwalbe, whose keen eye, good taste, and sense of structure improved the book in so many ways, and whose unflagging excitement about this project spurred me on when I needed it most.

Thanks to my family for their love and encouragement, and especially to my dad, who has—as always—been on my side, quietly cheering, smiling, urging me on. The incredible selflessness of my mother-in-law, Shirley Schiffman, made it possible for me to work for long stretches without feeling guilty about neglecting my baby girls. Thank you to my daughters: Leah, for bringing me back down to earth by being a toddler; and Joanna, for not being born too early or too late. My wife, Carole, has shown her love in countless ways—listening, reading, coaxing, forgiving, teaching me how to create, how to loosen up, how to let go. Her generosity of spirit gave me the freedom to be consumed by a sometimes needy, always present obsession.

Finally, thank you to the citizens of the United States for your trust and far-sightedness. By supporting the American research enterprise through agencies like the National Science Foundation, your taxes give scientists the most precious gift we could hope for—the chance to follow our imaginations wherever they may lead. I hope you take as much pleasure in our discoveries as we do.

I

LIVING SYNC

One

FIREFLIES AND THE
INEVITABILITY OF SYNC

---

*"Some twenty years ago I saw, or thought I saw, a
synchronal or simultaneous flashing of fireflies. I could hardly
believe my eyes, for such a thing to occur among insects is
certainly contrary to all natural laws."*

So wrote PHILIP LAURENT IN THE JOURNAL *Science* in 1917, as he joined the debate about this perplexing phenomenon. For 300 years, Western travelers to Southeast Asia had been returning with tales of enormous congregations of fireflies blinking on and off in unison, in displays that supposedly stretched for miles along the riverbanks. These anecdotal reports, often written in the romantic style favored by authors of travel books, provoked widespread disbelief. How could thousands of fireflies orchestrate their flashings so precisely and on such a vast scale? Now Laurent felt certain he had solved the enigma: "The apparent phenomenon was caused by the twitching or sudden lowering and raising of my eyelids. The insects had nothing whatsoever to do with it."

In the years between 1915 and 1935, *Science* published 20 other articles on this mysterious form of mass synchrony. Some dismissed the phenomenon as a fleeting coincidence. Others ascribed it to peculiar atmospheric conditions of exceptional humidity, calm, or darkness. A few believed there must be a maestro, a firefly that cues all the rest. As George Hudson wrote in 1918, "If it is desired to get a body of men to sing or play together in perfect rhythm they not only must have a leader but must be trained to follow such a leader. . . . Do these insects inherit a sense of rhythm more perfect than our own?" The naturalist Hugh Smith, who had lived in Thailand from 1923 to 1934 and witnessed the displays countless times, wrote in exasperation that "some of the published explanations are more remarkable than the phenomenon itself." But he confessed that he too was unable to offer any explanation.

For decades, no one could come up with a plausible theory. Even as late as 1961, Joy Adamson, in her sequel to *Born Free*, marveled at an African version of the same phenomenon, the first ever described on that continent:

> a great belt of light, some ten feet wide, formed by thousands upon thousands of fireflies whose green phosphorescence bridges the shoulder-high grass . . . The fluorescent band composed of these tiny organisms lights up and goes out with a precision that is perfectly synchronized, and one is left wondering what means of communication they possess which enables them to coordinate their shining as though controlled by a mechanical device.

By the late 1960s, the pieces of the puzzle began to fall into place. One clue was so obvious that nearly everyone missed it. Synchronous fireflies not only flash in unison—they flash in *rhythm*, at a constant tempo. Even when isolated from one another, they still keep to a steady beat. That implies that each insect must have its own means of keeping time, some sort of internal clock. This hypothetical oscillator is still unidentified anatomically but is presumed to be a cluster of neurons somewhere in the firefly's tiny brain. Much like the natural pacemaker in our hearts, the oscillator fires repetitively, generating an electrical rhythm that travels downstream to the firefly's lantern and ultimately triggers its periodic flash.

The second clue came from the work of the biologist John Buck, who did more than anyone else to make the study of synchronous fireflies scientifically respectable. In the mid-1960s, he and his wife, Elisabeth, traveled to Thailand for the first time, in hopes of seeing the spectacular displays for themselves. In an informal but revealing experiment, they captured scores of fireflies along the tidal rivers near Bangkok and released them in their darkened hotel room. The insects flitted about nervously, then gradually settled down all over the walls and ceiling, always spacing themselves at least 10 centimeters apart. At first they twinkled incoherently. As the Bucks watched in silent wonderment, pairs and then trios began to pulse in unison. Pockets of synchrony continued to emerge and grow, until as many as a dozen fireflies were blinking on and off in perfect concert.

These observations suggested that the fireflies must somehow be adjusting their rhythms in response to the flashes of others. To test that hypothesis directly, Buck and his colleagues later conducted laboratory studies where they flashed an artificial light at a firefly (to mimic the flash of another) and measured its response. They found that an individual firefly will shift the timing of its subsequent flashes in a consistent, predictable manner, and that the size and direction of the shift depend on when in the cycle the stimulus was received. For some species, the stimulus always advanced the firefly's rhythm, as if setting its clock ahead; for other species, the clock could be either delayed or advanced, depending on whether the firefly was just about to flash, whether it was halfway between flashes, and so on.

Taken together, the two clues suggested that the flash rhythm was regulated by an internal, resettable oscillator. And that immediately suggested a possible synchronization mechanism: In a congregation of flashing fireflies, every one is continually sending and receiving signals, shifting the rhythms of others and being shifted by them in turn. Out of the hubbub, sync somehow emerges spontaneously.

Thus we are led to entertain an explanation that seemed unthinkable just a few decades ago—the fireflies organize themselves. No maestro is required, and it doesn't matter what the weather is like. Sync occurs through mutual cuing, in the same way that an orchestra can keep perfect time without a conductor. What's counterintuitive here is that the insects don't need to be intelligent. They have all the ingredients they need: Each firefly contains an oscillator, a little metronome, whose timing adjusts automatically in response to the flashes of others. That's it.

Except for one thing. It's not at all obvious that the scenario can work. Can perfect synchrony emerge from a cacophony of thousands of mindless metronomes? In 1989 my colleague Rennie Mirollo and I proved that the answer is yes. Not only can it work—it will *always* work, under certain conditions.

For reasons we don't yet understand, the tendency to synchronize is one of the most pervasive drives in the universe, extending from atoms to animals, from people to planets. Female friends or coworkers who spend a great deal of time together often find that their menstrual periods tend to start around the same day. Sperm swimming side by side en route to the egg beat their tails in unison, in a primordial display of synchronized swimming. Sometimes sync can be pernicious: Epilepsy is caused by millions of brain cells discharging in pathological lockstep, causing the rhythmic convulsions associated with seizures. Even lifeless things can synchronize. The astounding coherence of a laser beam comes from trillions of atoms pulsing in concert, all emitting photons of the same phase and frequency. Over the course of millennia, the incessant effects of the tides have locked the moon's spin to its orbit. It now turns on its axis at precisely the same rate as it circles the earth, which is why we always see the man in the moon and never its dark side.

On the surface, these phenomena might seem unrelated. After all, the forces that synchronize brain cells have nothing to do with those in a laser. But at a deeper level, there is a connection, one that transcends the details of any particular mechanism. That connection is mathematics. All the examples are variations on the same mathematical theme: self-organization, the spontaneous emergence of order out of chaos. By studying simple models of fireflies and other self-organizing systems, scientists are beginning to unlock the secrets of this dazzling kind of order in the universe.

The question about self-organization that Rennie and I explored was originally posed by Charlie Peskin, an applied mathematician at New York University's Courant Institute. A soft-spoken man with a neatly trimmed beard and an easy smile, Peskin is one of the world's most creative mathematical biologists. He loves to use math and computers to plumb the mysteries of physiology: how the molecules and tissues and organs of the body perform their exquisite functions. Whether he's trying to work out how the retina can detect the dimmest light imaginable, or how molecular motors generate the forces in muscles, his trademark is his versatility. He seems willing to try anything, whatever is required to gain insight. If the math he needs does not exist, he'll invent it. If the problem requires a supercomputer, he'll program it. If existing procedures are too slow, he'll devise faster ones.

Even his mathematical style is flexible and pragmatic. His most celebrated work deals with the three-dimensional pattern of blood flow in the chambers of a pumping heart, complete with realistic anatomy, valves, and fiber architecture. For that complex problem he combined the brute force of a supercomputer simulation with the finesse of a wholly original numerical scheme. On other problems, however, he has usually followed Einstein's dictum that everything should be made as simple as possible, but not simpler. In those cases he opted for a minimalist approach, neglecting all biological details except the truly essential ones. It was in that minimalist spirit that Peskin proposed a schematic model for how the pacemaker cells of the heart might synchronize themselves.

The heart's natural pacemaker is a marvel of evolution, perhaps the most impressive oscillator ever created. A cluster of about 10,000 cells called the sinoatrial node, its function is to generate the electrical rhythm that commands the rest of the heart to beat, and it must do so reliably, minute after minute, for three billion beats in a lifetime. Unlike most of the cells in the heart, the pacemaker cells oscillate automatically—isolated in a petri dish, their voltage rises and falls in a regular rhythm.

All of which raises the issue, Why do we need so many of these cells, if one can do the job by itself? Probably because a single leader is not a robust design—a leader can malfunction or die. Instead, evolution has produced a more reliable, democratic system in which thousands of cells collectively set the pace. Of course, democracy raises its own problems: Somehow the cells have to coordinate their firings; if they send conflicting signals, the heart becomes deranged. And that's the issue that Peskin wondered about: How do these cells, with no leader or outside instructions, manage to get in sync?

Notice how similar this question is to the earlier one about fireflies. Both involve large populations of rhythmic individuals that fire off sudden pulses that jolt the rhythms of others in their group, speeding them up or slowing them down according to specific rules. In both cases, sync appears inevitable. The challenge is to explain why this should be so.

In 1975, Peskin examined this question within the framework of a simplified model. Each pacemaker cell is abstracted as an oscillating electrical circuit, equivalent to a capacitor in parallel with a resistor. (A capacitor is a device for storing electrical charge, and here plays a role akin to the cell's membrane; a resistor provides a pathway for current to flow out of the cell, analogous to so-called leakage channels in the membrane.) A constant input current causes the capacitor to charge up, increasing its voltage steadily. Meanwhile, as the voltage rises, the amount of current leaking through the resistor increases, so the rate of increase slows down. When the voltage reaches a threshold, the capacitor discharges, and the voltage drops instantly to zero—this pattern mimics the firing of a pacemaker cell and its subsequent return to baseline. Then the voltage starts rising again, and the cycle begins anew. Viewed as a function of time, the voltage cycle has two parts: a gentle ascent along a charging curve (a graph shaped like half an arch, rising but bowed downward), followed by a vertical drop back to baseline.

Next, Peskin idealized the cardiac pacemaker as an enormous collection of these mathematical oscillators. For simplicity, he assumed that all the oscillators are identical (and therefore follow the same charging curve); that each oscillator is coupled equally strongly to all the others; and that the oscillators affect one another only when they fire. Specifically, when an oscillator fires, it instantly kicks the voltages of all the others up by a fixed amount. If any cell's voltage exceeds the threshold, it fires immediately.

What makes the problem so bewildering is that different oscillators are typically at different stages in the cycle at any given moment—some are on the

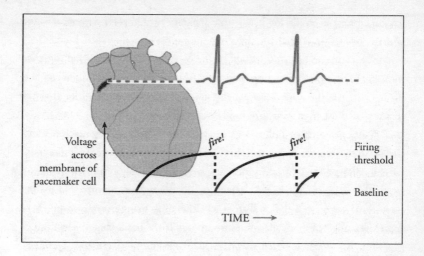

brink of firing, others are farther down on the charging curve, and still others may be close to baseline. Once the lead oscillator reaches threshold, it fires and kicks everyone else to different positions along the charging curve. The effects of the firing are mixed: Oscillators that were close to threshold are knocked closer to the firing oscillator, but those close to baseline are knocked farther out of phase. In other words, a single firing has synchronizing effects for some oscillators and desynchronizing effects for others. The long-term consequences of all these rearrangements are impossible to fathom by common sense alone.

For a more vivid picture of what's going on, imagine an individual cell as analogous to a toilet tank filling with water. As the water pours in, its level rises steadily, as the voltage does in the cell. Suppose that when the water reaches a certain height, the toilet automatically flushes. The sudden discharge returns the water to its baseline level, at which point the tank begins filling again, creating a spontaneous oscillator. (To complete the analogy, we also have to suppose that the tank is slightly leaky. Water spills out through a small hole near the bottom of the tank. It drains faster when there's more water in the tank, which implies that the tank fills more slowly as it rises. This leakage is not important for the oscillation itself—the apparatus would cycle without it—but it turns out to be crucial for the synchronization of many such oscillators.) Finally, imagine an army of 10,000 of these oscillating toilets, rigged together by a system of pipes connecting every tank to every other, so that when any one flushes, it raises the water level equally in all the rest. If that additional water lifts any of those over their threshold, they flush too.

It's a bizarre image, a plumber's version of a Rube Goldberg machine, and the question becomes, What will this contraption do, once started? Remain perpetually disorganized? Split into battling factions, flushing in turn?

Peskin conjectured that the system would always synchronize: No matter how it was started, all the oscillators would end up firing in unison. Furthermore, he suspected that sync would occur even if the oscillators were not quite identical. But when he tried to prove his conjectures, he ran into technical obstacles. There were no established mathematical procedures for handling large systems of oscillators coupled by sudden, discontinuous impulses. So he backed off and focused on the simplest possible case: two identical oscillators. Even here the mathematics was thorny. He restricted the problem further by allowing only infinitesimal kicks and infinitesimal leakage through the resistor. Now the problem became manageable; for this special case, he proved that sync was inevitable.

Peskin's proof relies on an idea introduced by the French mathematician Henri Poincaré, the founder of chaos theory. Poincaré's concept is the mathematical equivalent of strobe photography. Take two identical oscillators, A and B, and chart their evolution by taking a snapshot every time A fires. What does the series of snapshots look like? Oscillator A has just fired, so it always appears at baseline, at zero voltage. The voltage of B, in contrast, changes from one snapshot to the next. By solving the equations governing his model, Peskin found an explicit but messy formula for the change in B's voltage between snapshots. The formula revealed that if the voltage is less than a certain critical value, it will decrease steadily until it reaches zero, whereas if it is larger, it will increase steadily until it reaches threshold. In either case, B ends up synchronized with A. There is one exception: If B's voltage is precisely equal to the critical voltage, then it can be driven neither up nor down and so stays poised at criticality. The oscillators fire repeatedly half a cycle out of phase from each other. But this equilibrium is unstable: The slightest nudge tips the system toward synchrony.

Despite Peskin's successful analysis of the two-oscillator case, the case of an arbitrary number of oscillators eluded proof for another 15 years. During this time Peskin's work went virtually unnoticed. It lay buried in an obscure monograph—essentially a photocopied set of his lecture notes—available only by request from his department.

One day in 1989, I was flipping through a book called *The Geometry of Biological Time*, written by the theoretical biologist Art Winfree, one of my heroes. At the time I was a postdoctoral fellow in applied math at Harvard and feeling hungry for a new problem to work on. Even though I'd been poring over Winfree's book for the past eight years, I still found it to be an endless source of ideas and inspiration. It wasn't just a summary of past research on biological oscillators—it was a map for fortune hunters, a guide to future discoveries. On practically every page, Winfree pointed the way to good unsolved problems, with tips about which ones were ripest.

And here was a lead I hadn't noticed before: In a section on oscillators communicating by rhythmic impulses, Winfree mentioned the model for cardiac pacemaker cells that Peskin had proposed in his monograph. Although Peskin had successfully analyzed the case of two identical oscillators, wrote Winfree, "the population problem awaits completion."

That piqued my curiosity. What was this fundamental puzzle, all set up, waiting to be solved? I'd never heard of Peskin's work, but it sounded extraordinary. Nobody else had ever tried to tackle the mathematics of a population of "pulse-coupled" oscillators, where the interactions are mediated by abrupt, pulsatile signals. This was a noticeable hole in the literature of mathematical biology, and an embarrassing one at that, given how common it was for biological oscillators to interact in this way. Fireflies flash. Crickets chirp. Neurons spike. All use sudden pulses to communicate. Nevertheless, theorists shied away from pulse coupling for mathematical reasons. Impulses make variables jump discontinuously, and calculus has trouble coping with jumps—it works best for processes that change smoothly. Yet Peskin had somehow found a way to analyze two oscillators that repeatedly zap each other. How had he done it? And what blocked his path for more than two?

Our library didn't have a copy of his monograph, but Peskin kindly mailed me the relevant pages. His analysis was sweet, clear, and direct. But I quickly realized why he stopped at two: Although his analysis was elegant, his formulas were already becoming unwieldy. Three oscillators would be worse, and an arbitrary number, n, seemed downright forbidding. I couldn't see how to extend his argument or bypass the complications.

To get a better feel for the problem, I ran it on the computer in two different ways. The first approach was to inch ahead and try the three-oscillator problem, mimicking Peskin's strategy, using small kicks and leakage, and letting the computer handle all the algebra. The formulas were horrible—some of them filled several pages—but with the computer's help, I whittled them down to something intelligible. The results showed that Peskin's conjecture was probably true for three oscillators. They also showed that this was not the right way to proceed. The algebra, even with the help of the computer, was becoming prohibitive.

The second approach was simulation. No formulas now, just let the computer march the system forward in time, one small step after another, then see what happens. Simulation is no substitute for math—it could never provide a proof—but if Peskin's conjecture was false, this approach would save me a lot of time by revealing a counterexample. This sort of evidence is extremely valuable in math. When you're trying to prove something, it helps to know it's true. That gives you the confidence you need to keep searching for a rigorous proof.

Programming the simulation was easy. When one oscillator fires, it kicks all the others up by a certain, fixed amount. If any of the kicked ones go over the threshold themselves, let those fire too, and update the others accordingly. Otherwise, in between firings, use Peskin's formulas to advance all the oscillators toward their thresholds.

I tried a population of 100 identical oscillators. With their voltages initially scattered at random between baseline and threshold, I plotted them as a swarm of dots arching toward threshold, climbing up their common charging curve of voltage versus time. Even with the help of computer graphics, I couldn't see a pattern in their collective motion—only a buzzing confusion.

The problem here was too much information. And so I came to appreciate another advantage of Peskin's strobe method: Not only does it simplify the analysis, it's also the best way to visualize how the system evolves. All the oscillators are invisible except at the precise moments when one particular oscillator fires. At those moments, an imaginary strobe light illuminates the rest of the oscillators, revealing their instantaneous voltages. Then the whole system lapses back into darkness until the next time that distinguished oscillator fires. Peskin's model has the property that the oscillators fire in turn—no one ever jumps the queue—so 99 other oscillators fire in the dark before the next strobe flash occurs.

Viewed on the computer, these computations flew by so rapidly that the screen appeared to flicker, with 99 oscillators hopping along the charging curve, changing their positions with each flash of the strobe. Now the pattern was unmistakable. The dots clumped together, forming small pockets of sync that coalesced into larger ones, like raindrops merging on a windowpane.

It was spooky—the system was synchronizing itself. Defying Philip Laurent and all the other skeptics who had argued that firefly sync was impossible in principle, that such a thing was "certainly contrary to all natural laws," the computer was showing that a mob of mindless little oscillators *could* fall into step automatically. The effect was uncanny to watch. An onlooker couldn't help but feel that the oscillators were deliberately cooperating, consciously striving for order, but they were not. Each one was responding robotically to the impulses fired by others, with no goal in mind.

To make sure I hadn't gotten lucky on the first try, I repeated the simulation dozens of times, for other random initial conditions and for other numbers of oscillators. Sync every time. Peskin's conjecture seemed to be right. The challenge now was to prove it. Only an ironclad proof would demonstrate, in a way that no computer ever could, that sync was inevitable; and the best kind of proof would clarify *why* it was inevitable. I called my friend Rennie Mirollo, a mathematician at Boston College.

Rennie and I had known each other for ten years. As grad students at Harvard, we used to hang out together on weekends, eating french fries at greasy spoons at 2 A.M., while talking about math and women in roughly equal measure. But we never worked together in those days. His training was in pure math while mine was in applied math—we could understand each other, but not completely.

For his doctoral studies, Rennie worked on a very abstract problem and hoped to write his thesis about it. His instinct told him that a certain theorem must be true, and he spent three years trying to prove it. One day, he realized that it was false—he found a counterexample that wrecked everything. Nothing could be salvaged. Yet rather than be depressed, his reaction was to switch to a new branch of mathematics, solve a key problem in it, and write a thesis— all in one year.

Around 1987, Rennie and I began working together. Our strengths were complementary. Usually I would propose the problem, explain its scientific context, run computer simulations, and suggest intuitive arguments. He would come up with strategies to crack the problem wide open, and then find ways to prove a theorem.

When I told him about my computer experiments on Peskin's model, he was eager at first, good-natured and curious. But once he understood the question, he became impatient, like a boxer waiting to enter the ring. He gave me a few more minutes to summarize what I'd done, but before long, he insisted on looking at it his own way.

Rennie simplified the model ruthlessly. He had no patience for all the details inherent in Peskin's original circuit model, with its capacitors and resistors and voltages. The only essential feature of the model, he guessed, is that each oscillator follows a slowing upward curve of voltage as it rises toward threshold. So he imposed that geometry from the start. He threw away the circuit and replaced it with an abstract, voltagelike variable that repeatedly builds up to a threshold, fires, and resets. Then he imagined a collection of these variables, n of them, all identical, and all interacting as before: Whenever one oscillator fires, it pulls all the others up by a fixed amount, or up to threshold, whichever is less.

This distilled model is not only clearer (which reduces the algebra enormously); it's also more broadly applicable. Instead of a purely electrical interpretation in terms of voltage, we could now think of the variable as measuring any oscillator's readiness to fire, whether a heart cell or a cricket, a neuron or a firefly.

We were able to prove that this generalized system almost always becomes synchronized, for any number of oscillators and no matter how they are started. A key ingredient in the proof is the notion of "absorption"—a shorthand for the idea that if one oscillator kicks another over threshold, they will remain synchronized forever, as if one had absorbed the other. Absorptions were conspicuous in my computer experiments, when the oscillators appeared to merge like raindrops. They are also irrevocable: Once two oscillators fire together, they will never drift apart on their own, because they have identical dynamics; furthermore, they are identically coupled to all the others, so even when they are kicked, they will stay in sync because they are jolted equally. Thus absorptions act like a ratchet, always bringing the system closer to synchrony.

The heart of the proof is an argument demonstrating that a sequence of absorptions locks the oscillators together in ever-growing clumps, until they finally coalesce into one giant group. If you're not a mathematician, you might be wondering how to go about proving something like that. There are an infinite number of different ways to start the system, so how can all the possibilities be covered? And what ensures that enough absorptions will occur to carry the system all the way to ultimate synchrony?

As I outline the reasoning below, don't worry too much about following the details. The point is just to give you a sense of how such proofs are built. It's not like what you might expect if your only experience was with high school geometry, which is often taught in a mechanical, authoritarian way. Developing a mathematical proof is actually a very creative process, full of vague ideas and images, especially in the early stages. Rigor comes later. (If you are not particularly interested in this, feel free to skip ahead to page 30.)

The first step is to catalog all the possible starting configurations. For instance, let's reconsider the case of two oscillators. Because of Peskin's strobe trick, we know we don't need to watch the oscillators at all times. It's enough to focus on one moment in every cycle, which we choose to be the instant immediately after oscillator A has fired and returned to baseline. Then oscillator B could be at any "voltage" between baseline and threshold. Visualizing B's voltage as a point on a number line, with baseline at 0 and threshold at 1, we see there's a line segment of different possibilities. This one-dimensional segment encompasses all possible starting conditions for the system (because we know A is at 0, having just fired and reset to baseline; the only variable is B, which must be somewhere along the line segment between 0 and 1).

Three oscillators create a larger space of possibilities. Now we need to know two numbers: Given that A has just fired and returned to 0, we still need to specify the voltages of oscillators B and C at that instant. Visualize those two possibilities, all combinations of B's and C's voltages. What is the geometry corresponding to a pair of numbers? We can think of them as the two coordinates of a point in a two-dimensional space.

Picture the x, y plane, familiar from high school math. Here the x-axis, plotted horizontally as usual, represents B's voltage at the moment that A fires. The y-axis, plotted vertically, represents C's voltage at the same instant. A pair of voltages is a single point in this plane.

As we allow B and C to vary independently over all voltages between 0 and 1 (to cover all possibilities), the corresponding point moves around inside a square region, in the same way that turning the two knobs on an Etch A Sketch moves the mechanical pen across a square screen.

The upshot is that with three oscillators, we have a square of possible initial conditions: one axis for B, one for C. Notice that we don't need an axis for A, since it always starts at 0, by definition of how we strobe the system.

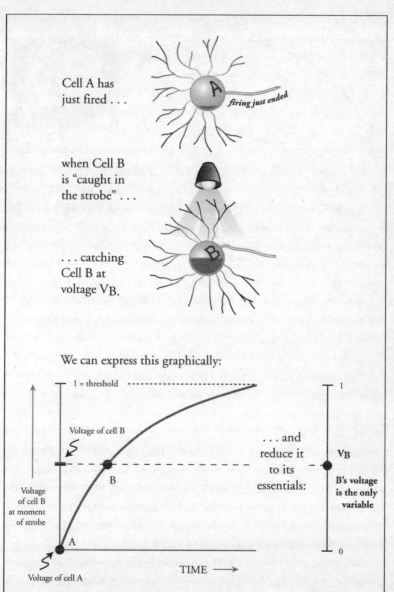

Cell A has just fired . . .

firing just ended

when Cell B is "caught in the strobe" . . .

. . . catching Cell B at voltage V_B.

We can express this graphically:

1 = threshold

Voltage of cell B

B

A

Voltage of cell B at moment of strobe

Voltage of cell A

TIME $\longrightarrow$

. . . and reduce it to its essentials:

V_B

B's voltage is the only variable

1

0

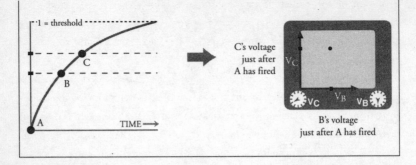

The pattern is becoming clear. As we add more oscillators, we need to add more dimensions to account for all the possibilities. Four oscillators require a solid cube of initial conditions; five require a four-dimensional hypercube; and in general, n oscillators require an $(n-1)$-dimensional hypercube. That sounds mind-boggling, and it is, if you try to picture it. But the mathematical formalism handles all dimensions in the same way. There are no new complications. So for concreteness, I'll continue to focus on the three-oscillator case, which contains all the main ideas.

The next step is to translate the dynamics—the evolution of the system in time—into the pictorial framework we're developing. The goal is to predict whether the system will end up in sync, given an initial condition for oscillators B and C.

Imagine what happens if we let the system run. All the oscillators rise toward threshold, fire, and reset to baseline; they also respond to kicks from other oscillators. To eliminate redundant information, we again exploit the strobe idea: Let the system run in the dark until the next time A has fired and gone back to 0, and B and C have responded. Then flash the strobe and take the next snapshot, recording the new positions of B and C.

The geometrical effect is that the old point in the square has just hopped to a new point: the updated voltages of B and C. In other words, the dynamical evolution of the system is tantamount to a transformation that takes any given point in the square and sends it to a new point, according to some complicated rule that depends on the shape of the charging curve and the size of the kicks.

The process can be repeated; the new point can be treated as an initial point, then sent on its way by the transformation, over and over again, hopping from place to place in the square in a series of jerky steps. If the system is destined to sync, the point will eventually hop toward the lower left corner of the square—the point with voltages (0,0)—meaning that each oscillator reaches baseline simultaneously. (Why that corner? Because that's where oscillator A is. By definition of the strobe, A has just fired and reset, so its voltage is 0. In the synchronized state, both of the other oscillators have voltage 0 too.)

In principle, every initial point has a fate that can be calculated. If all the oscillators end up firing in sync, we say the starting point is "good." Otherwise, it's "bad." Rennie and I never found a way to decide exactly which points were which, but we did manage to prove that almost all points are good. Bad points do exist, but they are so few and sparse that, taken all together, they occupy no area. Or to put it another way, if you choose a point at random, you have no chance of picking a bad one.

That might sound nonsensical: If bad points exist, you may be thinking, surely with my luck I would choose one. But you wouldn't. It would be like throwing a dart at a dartboard and requiring that it land precisely on the dividing line between two scores. That's unlikely enough, but now imagine that the line has no thickness (as required if it is to have zero area) and now you see why a random dart would never hit it.

It was Rennie's idea to think about the bad points, even though we were interested in the good ones. His strategy was reminiscent of the artist's concept of negative space: To understand the object, understand the space around it. In particular, he found a way to prove that the bad points occupy zero area.

To give the flavor of the argument, let's concentrate on the worst of the bad points, which I'll call the "terrible" ones. These are the most defiant in resisting the urge to sync; they never undergo *any* absorptions. When the system starts at a terrible point, no pair of oscillators (let alone the whole population) ever synchronizes.

To see why the terrible points have no area, think of them collectively as a set, and examine what happens when we apply the transformation to all the points in that set. Each terrible point will hop somewhere, but it will still be terrible after the transformation. That's almost a tautology: If a point never leads to an absorption, then after one iteration of the transformation, it still never leads to an absorption. Hence, the new point is terrible too. Since the original set included *all* terrible points (by definition), this new point must have been lurking somewhere in there to begin with.

The conclusion is that the transformed set lies entirely inside the original. In more visual terms, it's like those "before" and "after" pictures favored by advertisements for diet programs. The transformed set—the slimmed-down "after" picture—is contained entirely in the chubbier "before" picture, just as the diet promised.

So far the argument hasn't used any information about the shape of the charging curve or the size of the kicks. When we finally take those details into account, we come to what seems at first like a paradox, though it's actually the clincher for the argument. Rennie and I were able to prove that the transformation from "before" to "after" works somewhat like the enlarging function on a photocopier. Any set of points that you feed into the transformation comes out larger afterward, in the sense that its total area will be magnified by a factor larger than 1. It does not matter what set you choose (just as it doesn't matter what image you place on the photocopier); all sets get expanded in area. In particular, the terrible set expands. But wait—that means the terrible set becomes fatter, not skinnier, seemingly contradicting what we said above. To be more precise, the conundrum is that the transformed version of the terrible set has to sit inside the original, yet its area also has to get larger, which seems impossible.

The only way these two conclusions could be compatible is if the original set had zero area to begin with (the "before" picture must have been a stick figure). Then there's no contradiction—when multiplied by a number larger than 1, its area is still zero, so the transformed set can fit inside the original. And this is exactly what we wanted to show: The terrible points occupy no area. So you'll never choose them, if you pick an initial condition at random. Nor will you pick any other bad points. And that's why sync is inevitable for this model.

The same argument works for any number of oscillators, with the slight modification that area must be replaced by volume or hypervolume when there are four or more oscillators. In any case, the probability of starting at a bad point is always zero. Hence Peskin was right: In his model of identical, pulse-coupled oscillators, everyone ends up firing in unison.

In developing this proof, we found that Peskin's leakage assumption was crucial; otherwise the transformation from "before" to "after" doesn't expand area, and the whole argument breaks down. And in fact, it *has* to break down, because the theorem is false without that assumption. If the charging curve had bowed up instead of down—if the voltage accelerates up to threshold—our simulations showed the population doesn't necessarily synchronize. The oscillators can get stuck in a random-looking pattern of disorganized firing.

This delicate point often tripped up other mathematicians when I first gave lectures about our work; before I had a chance to explain it, some heckler (and usually there was one) would interrupt and say the theorem is trivial, that of course the oscillators will synchronize, since they're all identical and coupled equally to one another, and what else did I expect? But that objection is too facile—it overlooks the subtle influence of the charging curve's shape. Only when the curve bends in the right direction is sync inevitable. In biological terms, the shape of the charging curve determines whether kicks are more potent at the beginning of the cycle (near baseline) or at the end of the cycle (near threshold). When the curve bows downward as in Peskin's model, a given kick in voltage translates to a larger shift in phase for oscillators close to threshold, which in turn ensures that the system will synchronize, though seeing why requires a complicated calculation and is certainly not obvious.

Our proof of Peskin's conjecture was the first rigorous result about a population of oscillators coupled by sudden impulses. With regard to real fireflies or cardiac pacemaker cells, however, the model is plainly simplified. It assumes that the firing of one oscillator always kicks the others toward threshold, thereby advancing their phases; real biological oscillators can generally inflict both advances or delays. Moreover, the Thai fireflies that are most adept at synchronizing—a species known as *Pteroptyx malaccae*—use an altogether different strategy: They continually adjust their clocks' frequencies, not their phases, in response to incoming flashes. In effect they make them tick faster or slower, rather than pushing the minute hands forward or back. By further pretending that all oscillators are identical, the model neglects the genetic variability inherent in any real population. And finally, assuming that all oscillators affect one another equally is a crude approximation for heart cells, which primarily influence their nearest neighbors. Given all the limitations of our analysis, we were unprepared for the reaction it was about to provoke.

Within the next few years, more than 100 papers were written on pulse-coupled oscillators by scientists in disciplines ranging from neurobiology to geophysics. In neurobiology, theorists studying models of neural networks had grown impatient with the prevailing approach, in which neurons were described coarsely by their average rates of firing (the number of spikes per second) instead of in terms of the actual timing of the spikes themselves. The new framework of pulse-coupled oscillators fit perfectly with the needs and mood of the time.

By an accident of scientific sociology, or maybe because of a mysterious zeitgeist, in the early 1990s scientists in other fields were also thinking about these kinds of systems. For example, the influential Caltech biophysicist John Hopfield pointed out a connection between pulse-coupled neurons and earthquakes. In a simplified model of an earthquake, crustal plates continually pull on one another, building up stress until a threshold is crossed. Then the plates slip suddenly, releasing their pent-up energy in a burst. The whole process is reminiscent of the gradual rise and sudden firing of a neuron's voltage. In the earthquake model, the slippage of one plate may be enough to trigger others to slip (just as neural firing can set off a chain reaction of other discharges in the brain). These cascades of propagating events can give rise to earthquakes (or epileptic seizures). Depending on the exact configuration of the other elements of the system, the result may be a minor rumble or a massive quake.

The same mathematical structure cropped up in models of other interacting systems, ranging from forest fires to mass extinctions. In each case, an individual element is subjected to increasing pressure, builds up toward a threshold, then suddenly relieves its stress and spreads it to others, potentially triggering a domino effect. Models with this character were all the rage in early 1990s. The statistics of the cascades—most very small, but a few cataclysmic—were studied theoretically by the physicist Per Bak and his collaborators, in connection with what they called self-organized criticality.

Hopfield's insight was that self-organized criticality might be intimately linked to synchronization in pulse-coupled oscillator systems. The tantalizing possibility of a relationship between those two areas spawned dozens of papers exploring the possible ties. This episode exemplifies the ways that mathematics can expose the underlying unity of phenomena that otherwise seem unrelated.

Our work also attracted media attention, largely because of its connection to fireflies, which conjure up childhood memories of summer evenings spent catching the glowing insects in glass jars. As a result of this coverage, in 1992 I received a delightful letter from a woman in Knoxville, Tennessee, named Lynn Faust. In her gracious and unassuming way, she was about to shatter a myth about synchronous fireflies that had lasted for decades. She wrote:

> I am sure you are aware of this, but just in case, there is a type of group synchrony lightning bug inside the Great Smoky Mountain National Park near Elkmont, Tennessee. These bugs "start up" in mid June at around 10 pm nightly. They exhibit 6 seconds of total darkness; then in perfect synchrony,

thousands light up 6 rapid times in a 3 second period before *all* going dark for 6 more seconds.

We have a cabin in Elkmont (due to be destroyed by the Park Service in December 1992) and, as far as we know, it is only in this small area that this particular type of group synchronized lightning bug exists. It is beautiful.

These are very different from our regular lightning bugs that just seem to blink on and off anytime after dusk.

She went on to say that across the creek from the cabin, fireflies high on the hillside start their sequence a little bit ahead of those below, so light seems to ripple down "like a waterfall of fireflies."

She wrote to the Park Service, desperately worried that their plan to evict the Elkmont residents from their cabins could ruin the habitat before any scientists had a chance to study it. The spectacle was seen nowhere else in the park, not even a half mile away, which suggested to Lynn that the local residents must be doing something to enable it. She guessed that the key might be freshly mowed lawns: For 50 years, Elkmont residents had mowed their lawns roughly every two weeks. That allowed the firefly larvae to survive the winter by burrowing into the short, mossy grass. They also hatched there in the spring and bred there in the summer. In short, without the Elkmont residents around to mow the grass, she argued, the fireflies might be lost to science forever. In support of her lawn hypothesis, she noted that the highest concentrations of fireflies were found

> right up next to the cabins and extending out onto the mown lawn
> areas . . . no larvae have been located at Uncle Lem Owenby's former homeplace
> where regular mowing no longer occurs. In the 15 years that the forest has
> replaced lawn at Mayna McKinna's cabin way up Jake's Creek she has noticed a
> marked decrease in "her" firefly population.

Lynn was also driven by concerns over losing her cabin and community. The Faust family had enjoyed the light show for 40 years. Every June, three generations would wrap themselves in blankets and sit silently on their unlit porch, waiting for the entertainment to begin.

What was so familiar to the Fausts was new to science. These backyard observations were about to become the first well-documented case of synchronous fireflies in the Western Hemisphere. In the decades since the controversy erupted in *Science* magazine in the early 1900s, the dogma had been that the phenomenon never occurs here, only in Asia and Africa. I put Lynn in touch with Jonathan Copeland, a firefly researcher at nearby Georgia Southern University, who, along with his collaborator Andy Moiseff of University of Connecticut, confirmed that the fireflies at the Faust cabin were synchronous, lighting up within three-hundredths of a second of one another.

Although Elkmont was absorbed by the Great Smoky Mountain National Park in 1992, the fireflies have survived the change, and "The Light Show" has gone on to become a tourist attraction. As for Lynn Faust, she continues to be tuned in to the pervasiveness of sync in nature, and is still making her own discoveries. In a 1999 letter to me, she wrote: "Just another simple synchrony I noticed this spring—when 4 turkey gobblers (these were domestic) are together during the spring mating time they congregate in a circle and gobble in synchrony after (what appears to be) the head gobbler makes an initial gobble."

Not everyone is so appreciative of the wonders of synchrony in the animal world. On May 18, 1993, the tabloid *National Enquirer* ran an article titled "Govt. Blows Your Tax $$ to Study Fireflies in Borneo—Not a Bright Idea!" The piece mocked the National Science Foundation for funding one of Copeland's grant proposals, and reported that Representative Tom Petri, Republican from Wisconsin, "doesn't think the study is likely to be very illuminating—and he wants to squash it. 'Spending taxpayers' money studying fireflies doesn't sound like a very bright idea to me.'"

It's hard to blame Representative Petri for missing the point. The value of studying fireflies is endlessly surprising. For example, before 1994, Internet engineers were vexed by spontaneous pulsations in the traffic between computers called routers, until they realized that the machines were behaving like fireflies, exchanging periodic messages that were inadvertently synchronizing them. Once the cause was identified, it became clear how to relieve the congestion. Electrical engineers devised a decentralized architecture for clocking computer circuits more efficiently, by mimicking the fireflies' strategy for achieving synchrony at low cost and high reliability. (The humble creatures have even helped save human lives. Ironically, the same week that Representative Petri's quip appeared in the *Enquirer,* an article in *Time* magazine reported that doctors were borrowing the firefly's light-emitting enzyme, luciferase, to accelerate the testing of drugs against resistant strains of tuberculosis.)

Beyond serving as an inspiration to engineers, the group behavior of fireflies has broader significance for science as a whole. It represents one of the few tractable instances of a complex, self-organizing system, where millions of interactions occur simultaneously—when everyone changes the state of everyone else. Virtually all the major unsolved problems in science today have this intricate character. Consider the cascade of biochemical reactions in a single cell and their disruption when the cell turns cancerous; the booms and crashes of the stock market; the emergence of consciousness from the interplay of trillions of neurons in the brain; the origin of life from a meshwork of chemical reactions in the primordial soup. All these involve enormous numbers of players linked in complex webs. In every case, astonishing patterns emerge spontaneously. The richness of the world around us is due, in large part, to the miracle of self-organization.

Unfortunately, our minds are bad at grasping these kinds of problems. We're accustomed to thinking in terms of centralized control, clear chains of command, the straightforward logic of cause and effect. But in huge, interconnected systems, where every player ultimately affects every other, our standard ways of thinking fall apart. Simple pictures and verbal arguments are too feeble, too myopic. That's what plagues us in economics when we try to anticipate the effect of a tax cut or a change in interest rates, or in ecology, when a new pesticide backfires and produces dire, unintended consequences that propagate through the food chain.

The firefly problem poses many of the same conceptual challenges, though of course it's much easier than economics or ecology. We have a much better idea about the nature of the individuals (fireflies) and their behavior (rhythmic flashing) and their interactions (resetting in response to light) than we do about the global marketplace or ecological webs, with so many diverse companies and species and unknown modes of interaction among them. But it's still not easy. In fact, it's at the edge of what we understand today. As such, it's an ideal starting point for learning how math can help us unravel the secrets of spontaneous order, and a case study of what it can (and cannot) do for us at this primitive, thrillingly early stage of exploration.

Although synchrony is ubiquitous among living things, its function is not always obvious. Why, for instance, should fireflies flash in unison? Biologists have offered at least 10 plausible explanations. The oldest one is called the beacon hypothesis. It has been known for decades that only the males synchronize their flashes; so, according to this view, the light show is directed at the females—a collective invitation to come hither. By blinking in concert, the males reinforce that seductive signal, beaming it for miles through the jungle canopy, luring females who might not otherwise see any of them. This may be why synchrony is common in densely vegetated areas (like the jungles of Thailand and Malaysia, or the forest behind Lynn Faust's cabin) but rare in the open meadows of the eastern United States, where fireflies can easily tryst without it.

A second possible advantage of synchrony is that you might get lucky—a female with eyes for your look-alike neighbor might become confused and mate with you instead. For that matter, synchrony could be equally beneficial for confounding predators; it's always safest to blend in with a crowd. The latest theory is that synchrony reflects competition, not cooperation: Every firefly is trying to be the first to flash (because females seem to prefer that), but if everyone follows that strategy, sync automatically ensues.

For many other creatures as well, communal sync is somehow tied to reproduction. Periodical cicadas outwit their predators by hiding underground for 17 years; then millions of them burst out simultaneously in a monthlong mating frenzy and die. Groups of male fiddler crabs, each of which sports a single, comically huge claw, take best advantage of their natural endowments: They flirt with a female by surrounding her and waving their gigantic claws in unison. (The ritual looks like many maestros conducting a single musician.)

In our own species, it is the females who do the synchronizing. Most women are familiar with the phenomenon of menstrual synchrony, in which sisters, roommates, close friends, or coworkers find that their periods tend to start around the same time. Long dismissed as anecdotal, menstrual synchrony was first documented scientifically by Martha McClintock, then an undergraduate psychology major at Wellesley, an all-female college in Massachusetts. She studied 135 fellow students and had them keep records of their periods throughout the school year. In October, the cycles of close friends and roommates started an average of 8.5 days apart, but by March, the average spacing was down to 5 days, a statistically significant reduction. A control group of randomly matched pairs of women showed no such change.

There are various ideas about the mechanism of synchronization, but the best guess is that it has something to do with pheromones: unidentified, odorless chemicals that somehow convey a synchronizing signal. The first evidence for this came from an experiment reported in 1980 by the biologist Michael Russell. A colleague of his, Genevieve Switz, had noticed the effect in her own life; when rooming with a female friend during the summer, the friend's period would lock on to hers, then drift apart after they separated in the fall. This suggested that Genevieve was a powerful synchronizer. Russell tried to determine what it was about Genevieve that was so compelling. For the experiment, Genevieve wore small cotton pads under her arms and donated the accumulated sweat to Russell each day. He then mixed it with a little alcohol and dabbed this "essence of Genevieve" on the upper lip of female subjects, three times a week for four months.

The results were startling: After four months, the subjects' periods began an average of 3.4 days apart from Genevieve's, down from 9.3 days at the beginning of the experiment. In contrast, the cycles of a control group (whose upper lips were dabbed with alcohol only) showed no significant change. Evidently something in Genevieve's sweat conveyed information about the phase of her menstrual cycle, in such a way that it tended to entrain the cycles of the other women who got wind of it.

Later studies didn't turn out so neatly. Some found statistical evidence for synchrony, others did not. Skeptics have viewed the conflicting data as evidence of the weakness or coincidental nature of the phenomenon. Recent work by McClintock, now a biologist at the University of Chicago, suggests quite the opposite, that menstrual sync is only the most conspicuous consequence of a larger phenomenon: chemical communication between women. In a 1998 experiment, McClintock and her colleague Kathleen Stern found that if they took swabs from the armpits of women at different points in their menstrual cycles and dabbed them on the upper lips of other women, the donor secretions shifted the phase of the recipient's cycle in a systematic way. Swabs taken from women at the beginning of their cycles, in the follicular phase before ovulation, tended to shorten the cycles of the women who received them. In other words, the recipients ovulated several days earlier than they would have otherwise, based on their prior records. In contrast, swabs taken from women at the time of ovulation prolonged the cycles of the beneficiaries. And secretions collected in the luteal phase, in the days before menstruation, had no effect whatsoever.

The implication is that women in a close-knit group are always pushing and pulling on one another's cycles, unconsciously engaging in a silent conversation mediated by pheromones. One possible outcome is menstrual synchrony. But given that pheromonal signals can nudge cycles together or drive them apart, depending on when in the month the signals were produced, it should come as no surprise that synchrony is not inevitable here—asynchrony or even antisynchrony (with cycles diametrically opposed) should be possible, and indeed, they too have been observed.

The function of this chemical dialogue remains a mystery. It could be that women unconsciously strive to ovulate and conceive in step with their friends (to allow them to share child-rearing and breast-feeding duties) and to keep out of step with their enemies (to avoid competing with them for scarce resources). Far-fetched as it sounds, this scenario is known to occur with other mammals. Female rats in a synchronized group produce larger and healthier offspring than those reared by a solo mother. Reproductive sync has benefits for all if the other females in the group are cooperative.

From a mathematical perspective, McClintock's data confirm what you probably already guessed: As coupled oscillators, women are far subtler than fireflies. The biochemical push and pull between them does not always coerce them into synchrony, unlike the firefly species in Southeast Asia that synchronize their flashes all night long, every night of the year. The inescapable synchrony of those fireflies (and of cardiac pacemaker cells) is brutally inflexible, and for that reason, is rarely found in other biological settings. Like women, most oscillators sync in some circumstances and not in others.

So the model we considered earlier in this chapter is starting to look far too simple. Although it helped us understand how sync could be inevitable under certain conditions, it went too far—it didn't allow for anything else. A more refined theory of coupled oscillators should predict whether a particular group of oscillators will synchronize or not, and tell us what factors are decisive in that regard.

The theory should also allow for the full range of ways that oscillators interact. Recall that fireflies hit each other with sudden pulses—hammer blows of light—but then ignore one another during the rest of their cycles, whereas women grapple with one another's oscillators at all times. Both types of coupling are common in nature, but the existing model allows only for pulses. An improved model should accommodate continuous interaction as well.

Furthermore, we have assumed so far that all the oscillators in a given population are strictly identical. But real oscillators are always diverse, with a spectrum of natural cycle lengths. Just as one woman may menstruate on a roughly 25-day cycle while another goes 35 days between periods, all other kinds of biological oscillators display a statistical distribution of cycle lengths. Even electronic and mechanical oscillators that are manufactured to be nominally identical never really are, due to slight errors in fabrication, or variations in their material properties.

Unfortunately, these complications ratchet up the mathematical difficulties tremendously. It's one thing to wish for a more realistic model, and another to construct one that's tractable. No insight is gained if the model is as perplexing as the phenomena it's supposed to describe. This is what makes mathematical modeling an art as well as a science: An elegant model strikes just the right compromise between simplicity and verisimilitude. Today we have a beautiful model of sync that does precisely that. Its creation was a collective enterprise that spanned three decades, and required the efforts of three pioneers, the first of whom was one of the most visionary and eccentric thinkers of the twentieth century.

Two

BRAIN WAVES AND THE CONDITIONS FOR SYNC

N ORBERT WIENER WAS NEVER QUITE A CELEBRITY. But when his book *Cybernetics* appeared in the 1950s, it electrified the reading public. The reviewer for the *New York Times* called it "seminal . . . comparable in importance to Galileo or Malthus or Rousseau or Mill." Wiener proposed a unified framework for thinking about problems of communication and control, whether in nervous systems or societies, animals or machines, computers or people. It was more like a dream than a finished theory, and it turned out to be premature. Nobody today would say they work on cybernetics, but the first half of the word lives on as the trendy prefix in *cyberspace* and *cyberpunk*.

Among scientists, Norbert Wiener will never be forgotten, for reasons both serious and silly. On the serious side, his name is enshrined in the terminology of advanced mathematics: Wiener process, Paley-Wiener theorem, Wiener-Hopf technique, and so on. A former child prodigy who received his Ph.D. from Harvard at age 18, Wiener revolutionized the theory of random processes. His analysis of Brownian motion, the erratic jiggling of molecules in solution, went far beyond Albert Einstein's intuitive approach to the same problem, and his methods laid the foundation for Richard Feynman's work in quantum electrodynamics and for Fisher Black and Myron Scholes's Nobel Prize–winning work on finance.

On the silly side, mathematicians love to tell stories about Wiener. Short and spherical, with thick glasses and a penchant for smoking cigars, he could often be found riding his unicycle through the corridors of MIT. Even in a profession whose members are not known for their athleticism or common sense, Wiener stood out. After failing to return dozens of consecutive serves from his tennis partner, Wiener suggested that they switch rackets. He was so absentminded that when he and his family moved from Cambridge to Newton, his wife wrote out their new address and directions home from his office, knowing full well he would forget they had moved. Sure enough, Wiener used the note as scrap paper for some calculations, threw it away, and walked back to his old house. When he arrived, he realized he no longer lived there, so he stopped a little girl on the street and asked her if she knew where the Wieners had moved. She said, "Yes, Daddy, come with me."

Wiener is a central figure in the science of sync, in part because he asked a question that no one before him had dared to address. Whereas earlier mathematicians had been content to work on problems involving two coupled oscillators, Wiener tackled problems involving millions of them. Perhaps even more important, he was the first to point out the pervasiveness of sync in the universe. Chirping crickets, croaking frogs, flashing fireflies, gaps in the asteroid belt, generators in the power grid—Wiener spotted sync in all of them. Superficial differences did not distract him. He was looking for transcendent principles. And he thought he found one, while pondering the origin of human brain waves.

In the late 1950s, no one really knew why the brain should oscillate at all. But decades earlier, physiologists had discovered that if you attach two electrodes to different points on a person's scalp, there's a tiny voltage between them, and that voltage fluctuates in time. When the technology of electronic amplifiers became sufficiently well developed, these tiny electrical fluctuations, or "brain waves," could be conveniently displayed on a strip-chart recorder, where a small pen bobs up and down as a sheet of paper scrolls by. (The same technology is used in lie detector tests and heart monitoring, and should be familiar to anyone who has ever watched a television show that's set in a hospital.)

Electroencephalographers, the experts who measure brain waves, became very good at noticing characteristic patterns in these tracings of brain activity. One pattern, the so-called alpha rhythm, occurs in people who are awake but relaxed with their eyes closed. Subjectively, it feels like a pleasant, spacey state. On a strip chart, it looks like a prominent oscillation of roughly 10 cycles a second.

Wiener wanted to study the alpha rhythm in much finer detail, because he had a hunch about what its function might be: He thought it was the sound of the brain's master clock ticking. Just as a computer needs a clock to synchronize the passing of messages among its thousands of components, Wiener supposed that the brain would coordinate its myriad neural activities by forcing them all to march to the beat of a centralized drummer. Individual neurons could not possibly serve that purpose: They were known to be sloppy oscillators at best, too imprecise to function as clocks. Wiener hypothesized instead that the brain ingeniously builds an accurate clock from an enormous number of sloppy ones. Somewhere in the brain, he proposed, there might be millions of specialized oscillators, maybe individual neurons or small clusters of them, all discharging about 10 times a second. Like any other biological population, these oscillators were bound to be diverse: Some would be inherently faster than others, preferring to fire 12 times a second, while others might run slow, firing only 8 times a second, though most would be somewhere in the middle, with natural frequencies close to 10 cycles a second. Left to their own devices, this motley bunch of neural oscillators would fire off impulses at disparate rates, producing an electrical racket akin to the sound of an orchestra tuning up before a performance. To work together as an accurate clock, these hypothetical oscillators would need to cooperate, to sense one another's electrical rhythms and react accordingly so as to stay in step.

Wiener's notion was that the oscillators would spontaneously synchronize by pulling on one another's frequencies. If an oscillator was running too fast, the rest of the group would slow it down; if it was going too slowly, the others would speed it up.

To test whether this mechanism of frequency pulling actually operated in the brain, Wiener proposed looking for a telltale signature that it should leave on the alpha rhythm. An analogy with politics helps at this point. Think of the natural frequencies of the oscillators as being like the spectrum of political leanings in a hypothetical society. The most extreme left-wing radicals correspond to a tiny cohort of oscillators that would like to run at, say, 8 cycles per second. Inching to the right on the spectrum, we encounter a larger subpopulation of liberals at 9 cycles per second, a dominant core of centrists at 10, back down to a smaller group of conservatives at 11, and only a handful of right-wing zealots at 12 cycles per second. For simplicity, let's suppose that a graph of the number of people in each category follows the familiar bell curve, dominated by a powerful center, and tapering off symmetrically as we move out to the right and left wings.

Keep in mind that this picture shows innate tendencies only. These are the

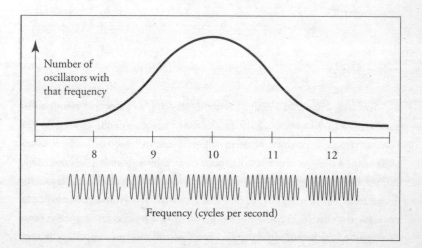

Number of oscillators with that frequency

8   9   10   11   12

Frequency (cycles per second)

attitudes that people would hold, or the frequencies that oscillators would exhibit, if they were completely shielded from the influences of others.

Now let individuals begin to pull on one another, and suppose (though politics rarely seems to work this way) that these oscillators can alter their frequencies. Through the persuasion of others, a slow oscillator can be convinced to run faster, and a faster one can be encouraged to slow down. Then, when the spectrum is measured, it will no longer resemble a bell curve. Wiener guessed that it would look something like this:

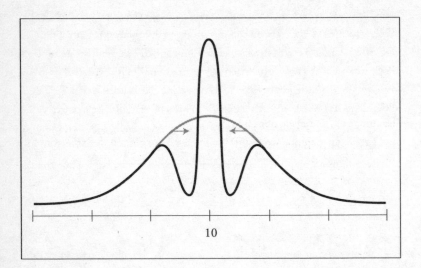

To make sense of this graph's peculiar shape, remember that most oscillators were near the middle of the bell curve to begin with. By pulling on one another's frequencies, many of them collapsed into the absolute center, forming a powerful mainstream consensus (the tall and narrow peak). Their combined influence on the rest of the population was strong enough to recruit a few moderates from either wing (increasing the height of the peak still further, and reducing the curve at the home positions of the moderates, causing the dips on both sides of the peak). Nevertheless, the consensus was not compelling enough to dislodge the most recalcitrant extremists on the fringes (shown as the shoulders on both ends of the spectrum).

Wiener predicted that the alpha rhythm would show this same peculiar peak and double dip in its spectrum of frequencies. If so, that would constitute strong evidence for his idea that the alpha rhythm is caused by synchronization between oscillators of diverse natural frequencies. To see if he was right, he would need a way to measure the spectrum with unprecedented precision. Here Wiener planned to exploit an experimental technique that his coworker Walter Rosenblith, an electrical engineer at MIT, had developed several years earlier. Rosenblith had found a way to record brain waves on magnetic tape, rather than on paper, which meant that the data could be processed electronically, yielding the first quantitative calculation of the brain wave spectrum. In contrast, all previous work was qualitative: it relied on pattern recognition, subjective judgments by trained human experts who had learned to spot patterns in brain wave squiggles. Now, with Rosenblith's approach, the calculation could be automated and made objective.

In a monograph he wrote in 1958, Wiener announced the results, though his presentation was suspiciously sketchy. Instead of showing the actual data (as any other scientist would have, if the findings were truly convincing), he drew a cartoon version of the measured spectrum, essentially the same as the diagram shown above. The results seemed a bit too pat, too good to be true. Wiener seemed to be hiding something.

Yet his writing betrayed no lack of confidence. He argued that frequency pulling was a universal mechanism of self-organization, operating not only on oscillators in the brain, but everywhere in nature, both among living things and nonliving ones. In an evangelical plea, he urged biologists to conduct experiments on frogs, crickets, and even the fireflies of Southeast Asia, long before the reality of their synchronous flashing had been established in the scientific literature. "Without daring to pronounce on the outcome of experiments which have not been made, this line of research strikes me as promising and not too difficult," he wrote in 1961.

His next task was to hammer out a detailed theory of frequency pulling.

Unfortunately, when he tried to back up his intuition with rigorous mathematics, he ran into insurmountable difficulties. He did present some rough calculations, but they were awkward and led nowhere. Wiener died in 1964 without having solved his pet problem. A year later, a college student would discover the right way to approach it.

At the time, Art Winfree was a senior majoring in engineering physics at Cornell. He had long dreamed of becoming a biologist, but instead of choosing the conventional route, he opted for hard-core training in math and physics, hoping to acquire a different set of tools. Electronics and computers, quantum mechanics and differential equations: these were things that most biologists never picked up.

When Winfree thought about the problem of group sync, he thought about the oscillators themselves, not merely their frequencies. In this respect, his conceptualization of the problem was much more explicit than Wiener's. He didn't just label each oscillator by how fast it tended to run (its location on the political spectrum, in the earlier analogy). Instead, he pictured it running step by step through its cycle, which is, after all, the quintessential thing that every oscillator does. Right away, that raised complications that would have repulsed nearly anyone else. But that's the advantage of youth—you don't know what's impossible.

His model was deliberately broad-brush. He intended it to be general enough to apply to *any* population of biological oscillators. The only way to capture the common features of chorusing crickets, flashing fireflies, pulsing pacemaker neurons, and the like was to ignore all their biochemical differences and to focus instead on the two things that all biological oscillators share: the ability to send and receive signals.

What makes the problem so confusing is that both of these properties change throughout an oscillator's cycle; influence and sensitivity are both functions of phase. For instance, a firefly's cycle consists of a sudden flash, then an interval of darkness while it is recharging its flash organ, then another flash, and so on. Experiments have shown that fireflies on the receiving end take heed of another's flash, and ignore the darkness. So in Winfree's mathematical description, the "influence function" would vary between two levels: large during the flash portion of the cycle, and nearly zero during the darkness. Similarly, a "sensitivity function" encodes how an oscillator responds to the signals it receives. Seeing a flash during one part of its cycle might cause a firefly to speed up its internal timer. Seen at another time, the same stimulus might slow the timer down, or have no effect at all. Those two functions were all Winfree needed to characterize an oscillator in his model. Once they were selected, the oscillator's behavior was determined, both as a sender and a receiver of signals. .

To make these ideas as concrete as possible, picture an oscillator as being like a jogger running around a circular track. The different locations on the track represent different phases in the oscillator's cycle of biological activity. For example, if the track represents the menstrual cycle, one place would correspond to ovulation. Another, halfway around the track, would correspond to menstruation, with in-between places corresponding to intermediate hormonal events. After one lap, the runner is back at ovulation again. Or if the track is supposed to represent the flash rhythm of a firefly, different locations would signify the flash itself followed by the various stages of biochemical recovery as the insect's flash organ recharges and builds up to its next firing.

In this way of thinking, two coupled oscillators are like joggers that continually shout instructions to each other as they run. The things they shout, and how loud they shout them, are determined by their current locations on the track; this information is encapsulated in Winfree's influence function. For example, if the value of one runner's influence function is currently small and positive, he shouts to the other one, "Hey, please go a little bit faster." On the other hand, a large negative value for the influence function means "You're going much too fast—slow down!" And a zero value of influence means the runner says nothing at all to his partner. As time passes, both runners advance around the track, so the instructions they shout keep changing from moment to moment.

This framework is extremely general. It can accommodate the pulselike interactions used by fireflies, crickets, and neurons (analogous to a sudden shriek, followed by silence for the rest of the cycle), or the ongoing push and pull of pheromones discovered by McClintock and Stern for the menstrual cycle (an ever-changing series of requests to speed up or slow down).

Meanwhile, both runners listen as well as shout. How they react to an incoming message is determined by Winfree's other function, the sensitivity function, which also varies from place to place along the track. When sensitivity happens to be high and positive, a runner is compliant and will follow whatever directions are coming at him at that moment. If sensitivity is zero, he ignores the instructions. And if sensitivity is negative, he is contrary: He speeds up when told to slow down, and vice versa. Here too, the model is very general, much more so than the Peskin model discussed in the last chapter, which assumed that oscillators always advance when kicked by a pulse. In Winfree's model, oscillators can advance *or* delay, depending on where they are in their cycle when they receive a pulse. Experiments have shown this is how most real biological oscillators behave.

For simplicity, Winfree further assumed that all the oscillators in a given population have the same influence and sensitivity functions. But he did allow for diversity in the same way that Wiener did before him: He assumed that the natural frequencies of the oscillators were randomly distributed across the population, according to a bell-shaped curve. In terms of the track analogy, you should visualize this population of oscillators as a running club with thousands of members on the track at the same time. Most of the runners are of average speed, but the club also includes some fast guys, former track stars in high school, and some slowpokes, trying to get back in shape after years of sloth. In other words, there is a distribution of natural abilities of the runners in the club, just as there is a distribution of natural frequencies of the oscillators in the biological population.

As if all this weren't complicated enough, there's one final aspect of the model that still needs to be specified: the connectivity. Winfree had to make an assumption about who is shouting at whom, and who is listening to whom. That would vary a lot, depending on what biological example he had in mind. Take circadian (roughly 24 hour) rhythms: In that case, Winfree guessed there might be clock cells all over the body, each secreting chemicals into the bloodstream on a daily cycle. Every cell would be bathed in the combined secretions of all the others; in effect, every cell communicates with every other. On the other hand, crickets pay most attention to the chirps coming from their immediate neighbors. And for oscillating neurons in the brain, the tangle of interconnections was unfathomable.

Winfree sidestepped these questions of connectivity and cut to the simplest problem, recognizing that it would still be fiendishly difficult. What would happen, he wondered, if each oscillator were influenced *equally* by all the others? It would be as if each runner were responding to the combined shouting of all the others, rather than just to the people running near him. Or to use a more realistic analogy, imagine sitting in a crowded concert hall after a magnificent performance. If the audience starts to clap in unison, you will be driven by the thunderous rhythm of the whole room, rather than by the couple sitting next to you.

Winfree wrote equations for his system of oscillators, describing how fast each one moves through its cycle. At any instant, an oscillator's speed is determined by three contributions: its preferred pace, which is proportional to its natural frequency; its current sensitivity to any incoming influences (which depends on where it is in its cycle); and the total influence exerted by all the other oscillators (which depends on where they all are in their cycles). It's a tremendous amount of mathematical bookkeeping, but in principle, the behavior of the entire system for all time is determined by the current locations of all the oscillators. In other words, complete knowledge of the present enables complete prediction of the future—at least in principle.

The calculation proceeds methodically. Given the locations of all the oscillators, we can compute their instantaneous speeds from Winfree's equations. Those speeds then tell us how far everyone will advance in the next instant.

(Pretend that an instant is just a very short time interval, and that all the oscillators move steadily during that time. Then the distance each oscillator travels around the circle equals its speed times the duration of the trip, just as it would for cars cruising down a highway.) So all the oscillators can now be advanced to their new phases, and the calculation is repeated, over and over again, marching forward an instant at a time. Conceptually at least, if we iterate this process long enough, we will see what fate holds for this community of oscillators.

What I've just described is called a system of differential equations. Such equations arise whenever we have rules for speeds depending on current positions. Problems like this have been studied since the time of Isaac Newton, originally in connection with the motion of planets in the solar system. There, each planet pulls on all the others by gravity, changing their locations, which in turn changes the gravitational forces between them, and so on—a hall of mirrors much like Winfree's oscillators with their ever-changing phases and forces of influence and sensitivity. It was precisely to solve baffling problems like this that Newton invented calculus. In one of the great achievements of Western science, he solved the "two-body problem" and proved that the orbit of the earth around the sun was an ellipse, just as Kepler had claimed before him. Curiously, however, the three-body problem turned out to be utterly intractable. For two centuries, the world's best mathematicians and physicists tried to find formulas for the motions of three mutually gravitating planets, until the late 1800s when the French mathematician Henri Poincaré proved that the task was futile. No such formulas could exist.

Since then, we've come to realize that most systems of differential equations are unsolvable, in that same sense; it's impossible to find a formula for the answer. There is, however, one spectacular exception. Linear differential equations are solvable. The technical meaning of *linear* need not concern us just yet; what matters is that linear equations are inherently modular. That is, a big, messy linear problem can always be broken into smaller, more manageable parts. Then each part can be solved separately, and all the little answers can be recombined to solve the bigger problem. So it's literally true that in a linear problem, the whole is exactly equal to the sum of the parts.

The hitch, though, is that linear systems are incapable of rich behavior. The spread of infectious diseases, the intense coherence of a laser beam, the roiling motion of a turbulent fluid: All of these are governed by *nonlinear* equations. Whenever the whole is different from the sum of the parts—whenever there's cooperation or competition going on—the governing equations must be nonlinear.

So it was hardly surprising that when Winfree looked at his differential equations for biological oscillators, he saw they were nonlinear. All the linear techniques he had learned in his physics and engineering classes were of no use to him now; he would never be able to find formulas for this problem. And as for nonlinear techniques, the few that were available were restricted to very small systems, like a single oscillator or two coupled oscillators. For the kind of question he was asking, about the population dynamics of thousands of interacting nonlinear oscillators, he would have to find his own way.

Winfree used a computer to simulate his model. Instead of math, it was more like doing an experiment. The computer would keep track of the oscillators as they ran around the circle at their variable speeds. The machine didn't care about linear or nonlinear, formulas or no formulas. It would just chug along, marching forward by one small step at a time, providing a good approximation to the model's true behavior. Winfree hoped the results might give him some intuition about how the oscillators would behave. At least he could see what would happen, even if he could not quite understand why.

Actually, one limiting case is easy to understand. If the oscillators completely ignore one another, they diffuse all over the circular track, because every one runs at its preferred speed, unaffected by the others. The faster ones overtake the slower ones and eventually lap them. In the long run, there are oscillators everywhere. Such a system is said to be incoherent. It's like the way that

American concert audiences applaud. We tend to ignore one another, and clap at whatever rate feels personally comfortable. The overall effect is a steady, arrhythmic clamor.

Winfree's simulations often settled into this same sort of incoherence, even when the oscillators *were* allowed to influence one another. For various combinations of sensitivity and influence functions, the population actively opposed synchronization. Even if all the oscillators were started in phase, they bucked the conformity and disorganized themselves. The population insisted on anarchy.

But for other influence and sensitivity pairs, Winfree found that the population would spontaneously synchronize. No matter how the oscillators were phased initially, some of them always congealed into a tight clump and ran around the circle together. Now the population acted more like an Eastern European concert audience, in which synchronized applause bursts out without any prompting.

In cases like this, synchronization occurred cooperatively. Once a few oscillators happened to sync by chance, their combined, coherent shouting stood out above the background din, and exerted a stronger effect on all the others. This nucleus recruited other oscillators toward them, which made the nucleus even larger and amplified its signal. The resulting positive feedback process led to a runaway, accelerating outbreak of synchrony, in which many oscillators rushed to join the emerging consensus. Some oscillators nonetheless remained unsynchronized because their natural frequencies were too extreme for the coupling to pull them in. The end result was a population split into a synchronized pack and a disorganized band of fringe oscillators.

When the system was self-synchronizing, Winfree found that no oscillator was indispensable. There was no boss. Any oscillator could be removed and the process would still work. Furthermore, the pack did not necessarily run at the speed of its fastest member. Depending on the choice of influence and sensitivity functions, the group could run at a pace nearer to the average speed of those in the pack, or it could go faster or slower than any of its members. It was all wonderfully counterintuitive. Group synchronization was not hierarchical, but it wasn't always purely democratic either.

Winfree's most important discovery came as a result of a strange and truly imaginative thought-experiment. Instead of a single population of oscillators, characterized by a single bell curve of natural frequencies, he imagined a family of such populations, each more homogeneous than the preceding one. Or in terms of our analogy, imagine many different running clubs.

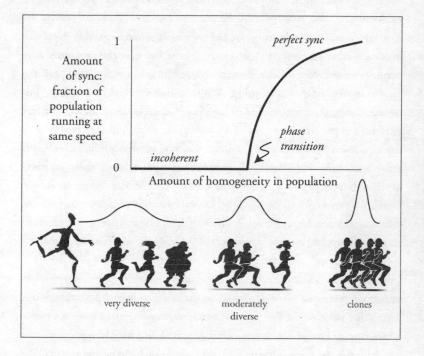

The first is extremely diverse, with members ranging widely in ability. He found that a club like this could never sync. None of its members could ever run as a pack, even if their influence and sensitivity functions predisposed them to do so. They would end up shouting and listening in vain; their heterogeneity would overwhelm their mutual desire to run together and scatter them all over the circle, just as if they were ignoring one another and running at their preferred speeds.

Now examine a club that is similar to the first but slightly more uniform. Its members have the same influence and sensitivity functions, but their abilities fall on a narrower and taller bell curve (meaning that more runners are just average, with fewer extreme slowpokes or track stars). You would think that this club should stand a better chance of synchronizing, at least in part, but Winfree found otherwise. As he considered increasingly homogeneous populations of oscillators, no sync occurred until he reached a critical point, a threshold of diversity. Then, suddenly, some of the oscillators spontaneously locked their frequencies and ran around together. As he made the distribution even narrower, more and more oscillators were co-opted into the synchronized pack.

In developing this description, Winfree discovered an unexpected link between biology and physics. He realized that mutual synchronization is analogous to a phase transition, like the freezing of water into ice. Think for a moment about how astonishing the phenomenon of freezing really is. When the temperature is just 1 degree above the freezing point, water molecules roam freely, colliding and tumbling over one another. At that temperature, water is a liquid. But now cool it ever so slightly below the freezing point and suddenly, as if by magic, a new form of matter is born. Trillions of molecules spontaneously snap into formation, creating a rigid lattice, the solid crystal we call ice. Similarly, sync occurs abruptly, not gradually, as the width of the frequency distribution is lowered through the critical value. In this analogy, the width of the distribution is akin to temperature, and the oscillators are like water molecules. The main difference is that when the oscillators freeze into sync, they line up in time, not space. Seeing that conceptual switch was a creative part of Winfree's analogy.

With this discovery, Winfree forged a connection between two great bodies of thought that had rarely noticed each other in the past. One was nonlinear dynamics, the study of the complex ways that systems can evolve over time; the other was statistical mechanics, the branch of physics that deals with the collective behavior of enormous systems of atoms, molecules, or other simple units. Each subject had strengths that complemented the other's weaknesses. Nonlinear dynamics worked well for small systems with only a handful of variables, but it couldn't handle the large constellations of particles that were child's play for statistical mechanics. On the other hand, statistical mechanics was wonderful for analyzing systems that had relaxed to equilibrium, but it couldn't cope with the incessant ups and downs of anything that oscillated or otherwise kept changing in time.

Winfree had now paved the way to a hybrid theory, which promised to be far more powerful than either one separately. This was to be a crucial step in the development of a science that could finally contend with the mysteries of spontaneous order in time as well as in space. And at a more practical level, it meant that the analytical techniques of statistical physics could now be brought to bear on the puzzle of how brain cells, fireflies, and other living things manage to synchronize with one another.

A few years later, a young Japanese physicist named Yoshiki Kuramoto learned of Winfree's work. He too was fascinated by self-organization in time, and he wanted to find a way to penetrate to its mathematical core. In 1975 he focused on a simpler, more abstract version of Winfree's model, and in a dazzling display of ingenuity, he showed how to solve it exactly.

This was a stunning achievement. Here was a system of infinitely many differential equations, all nonlinear, all coupled together. Such things are hardly ever solvable. The few exceptions that do exist are like diamonds, prized for their beauty, and for the rare glimpse they provide of the inner facets of nonlinearity. In this case, Kuramoto's analysis revealed the essence of group synchronization.

At first glance, it's hard to see what's so special about the structure of Kuramoto's model. As in Wiener's work, it still describes a huge population of oscillators with a bell-shaped distribution of natural frequencies; as in Winfree's model, every oscillator interacts equally with every other. Kuramoto's key innovation was to replace Winfree's influence and sensitivity formulation with a special kind of interaction, a highly symmetrical rule that embodies and refines Wiener's concept of frequency pulling.

The nature of the interaction is easiest to understand for a population of just two oscillators. Picture them as friends jogging together on a circular track. Being friends, they want to chat as they jog, so each makes adjustments to his preferred speed. Kuramoto's rule is that the leading one slows down a bit, while the trailing one speeds up by the same amount. (To be more precise, the amount of the adjustment is given by the sine function of the angle between them, multiplied by a number called the coupling strength, which determines the maximum possible adjustment.) This corrective action tends to synchronize the oscillators. Still, if the difference in their natural speeds is too large compared with the coupling strength, they won't be able to compensate for their different abilities. The faster one will gradually drift away from the slower one and lap him, in which case they should both think about finding new jogging partners.

What makes this rule so mathematically obliging is its symmetry. There are no distinguished places on the track, unlike in Winfree's original formulation, where different locations correspond to different salient events in a biological cycle of activity. For Kuramoto, all locations are indistinguishable. There are no landmarks. In effect, the runners have no way of knowing where they are, so they run in silence—no shouting or listening anymore—but they do watch each other carefully. Wherever they are on the track, they make the appropriate adjustments to their speeds, using a formula that depends only on the separation between them, not on where they happen to be.

Now imagine a much larger population of oscillators, and as before, picture it as a running club with members of diverse abilities. The interaction rule is that each runner looks at all the others, computes a tentative velocity correction relative to each, and averages them all to obtain the correction that will actually be made. For instance, suppose the runners happen to form a fairly tight pack at some moment. Kuramoto's rule tells the leader to slow down from his preferred speed, a sensible thing since everyone is behind him. A runner in the middle of the pack receives mixed messages, some telling him to speed up, others to slow down. A runner at the back feels the peer pressure to go faster.

All these corrections are happening instant by instant, oscillator by oscillator. To make the problem of coordinating themselves interesting, suppose the runners agree to start at random places on the track. There's no pack initially.

Even if a pack forms, it will not necessarily be arranged with the fastest ones in front; any arrangement is possible. The pack will keep shifting shape all the while, changing leaders, as the runners sort themselves out.

It's not obvious what will happen after a long time. The track stars may peel off and begin lapping the main pack, while the dawdlers fall behind. Or there may not even be a pack. The range of speeds may be so broad that the whole club falls apart, causing runners to diffuse all over the circle. In that case, everyone receives such mixed messages—go faster, go slower—that the velocity corrections cancel out, leaving everyone to run at his own favorite pace.

In his analysis of this confusing situation, Kuramoto found it helpful to quantify the degree of synchronization with a single number called the order parameter.

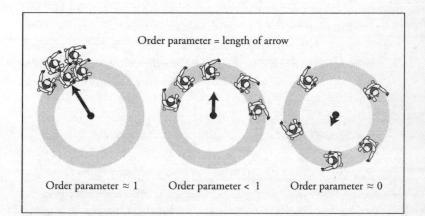

Order parameter = length of arrow

Order parameter ≈ 1   Order parameter < 1   Order parameter ≈ 0

Intuitively, everyone running shoulder to shoulder is a tighter form of sync than if the pack is spread out, and so should receive a higher sync score, a higher order parameter. The numerical value of the order parameter is always somewhere between 0 and 1, and is calculated from a mathematical formula that depends on everyone's relative position. At one extreme, when everyone is in perfect sync, running in unison, the order parameter equals 1. A looser pack has order parameter less than 1. At the opposite extreme, with runners scattered randomly all over the track, the order parameter equals 0.

Unlike Winfree, Kuramoto did not use a computer to provide hints about how the system would behave. He was guided by intuition alone. That makes his guess about the eventual outcome even more prescient: Kuramoto predicted that in the long run, the population would always settle into a state that's as steady as possible. The runners are still going, but their relative positions in the pack are not changing, so the order parameter is constant. Furthermore, the pack itself coasts at some compromise speed determined by its members. Kuramoto guessed this speed should be constant too.

In a daring mathematical leap, Kuramoto sought only those solutions of his equations that matched his intuition. If a solution did not have a constant order parameter and constant pack speed, he wasn't interested in it. He knew what he was looking for, and he was going to ignore everything else. It was a bold way to reason, because if the truth lay elsewhere, he would miss it. The other danger was that he might come up empty-handed; there might not be any solutions of the type he desired. Nevertheless, he guessed that there were, and he set out to find them. To give himself as much flexibility as possible, he did not specify in advance what the values of the order parameter or the pack speed must be, only that they be constant. Determining their values was part of the problem.

He found that the system could satisfy his demands in two very different ways. The order parameter could equal 0 forever, meaning that the population is totally and permanently disorganized. No pack ever forms. You'd see runners of all speeds everywhere on the track. This is as far from sync as the system could be. Surprisingly, this "incoherent state" is always a possible outcome, no matter how diverse or similar the runners are. Even if they are identical, incoherence can persist forever, once initially arranged. The intuition is that the runners have nothing to latch on to, no pack to draw them in, so by default each person runs at his natural speed and the whole population remains as disorderly as before. The other possibility is a "partially synchronized" state consisting of three groups: a synchronized pack of average runners; a slower, desynchronized swarm of dawdlers; and a faster, desynchronized swarm of sprinters. Unlike incoherence, this state is *not* always possible. Kuramoto found that it exists only up to a certain threshold of diversity. If the bell curve is broader than this threshold, meaning that the club membership is too diverse, the partially synchronized state disappears. The implication is that in a population of fireflies or brain cells, the oscillators have to be similar enough or nobody will synchronize at all.

In one stroke, Kuramoto had vindicated both Wiener and Winfree. The partially synchronized state is exactly what Wiener had in mind when he modeled the alpha rhythm of brain waves. The narrow peak at the center of Wiener's spectrum corresponds to the synchronized pack, and the tails on either side correspond to the desynchronized oscillators that are too fast or slow to be recruited. The phase transition that Winfree had discovered was the same as the threshold that Kuramoto was now finding. As they both realized, a synchronized pack cannot form unless the population is homogeneous enough. Wiener had missed that important point.

Beyond seeing the phase transition, what was new and delightful was that Kuramoto could derive an exact formula for it. Furthermore, he could calculate precisely how ordered the pack would be, as a function of the width of the bell curve. His formulas showed that a tiny synchronized nucleus is born at threshold, with order parameter barely above 0. As the diversity is reduced and the oscillators become more similar, the order parameter rises as the synchronized pack conscripts more of the population. Finally, at a width of absolute zero (corresponding to identical oscillators), Kuramoto's formula predicts a state of perfect order, with everyone in sync.

Soon after I finished my Ph.D. in 1986, I began a postdoctoral fellowship with Nancy Kopell, a mathematician at Boston University. Nancy was in her early forties, just entering the prime of her career. Attractive, funny, an incisive thinker and engaging lecturer, she was starting to be recognized as one of the best mathematical biologists around. (In particular, she and her collaborator

Bard Ermentrout were creating a stir by bringing new mathematical techniques to the study of the nervous system.) We'd met a few times at conferences, and she seemed like an ideal mentor for this next stage in my career, when my goal was to deepen my training in mathematics. When I mentioned that I'd like to work on a problem about populations of oscillators, she suggested that I look into Kuramoto's model.

I was instantly infatuated with it. In my graduate school courses, we were always taught that large nonlinear systems were monsters, practically impossible to solve. Yet here was one, and it was beautiful. It didn't even seem that hard to understand. As I read through Kuramoto's argument, I felt like I was following him, line by line. Nancy smiled at my enthusiasm, and then gently pointed out all the soft spots in Kuramoto's argument, all the unjustified leaps of logic. There were plenty of opportunities here for a budding mathematician. My job would be to put Kuramoto's intuitions on a firmer footing. I worked with Nancy for the next year, trying to prove a theorem that we were both convinced must be true. Though I never managed to solve that problem, I did find myself growing obsessed with the model.

Even after my postdoctoral fellowship was over, I continued to think about the model, off and on, for the next several years. The aspect that enchanted me had to do with the emergence of order out of randomness. How can a system of millions of particles spontaneously organize itself? The question sounds mystical, with religious overtones reminiscent of the story of creation in the Bible, where in the beginning the Earth was unformed and void: a condition that the ancient Greeks called chaos.

We may never understand the origins of order in the real universe, but in the imaginary universe of the Kuramoto model, the problem simplifies so much that we can address it mathematically. Here the genesis question becomes, How does incoherence give birth to synchrony? It dawned on me one day that there was a straightforward way to frame the question as an exercise in differential equations: I needed to view incoherence as an equilibrium state and then calculate its stability.

To clarify the mathematical meanings of those familiar words, *equilibrium* and *stability*, consider some examples from around the house. Imagine placing a glass of water on the kitchen table. For a second or two the water sloshes around in the glass, then comes to rest. Now the water surface is flat and horizontal. This is an equilibrium state, in the sense that the water will stay like that indefinitely. The equilibrium is additionally said to be stable because if we shake the glass a little and then stop, the water surface will return to level. Thus, equilibrium means nothing changes; stability means slight disturbances die out. Now try another example. Take a pencil and sharpen it, then stand it upright and carefully balance it on its point. Let go. If the pencil is poised perfectly, it will continue standing upright, so by definition this is also an equilibrium state. But obviously it's unstable: The slightest breeze will tip the pencil over and it won't re-right itself.

For the Kuramoto model, incoherence is an equilibrium state; if the oscillators of each frequency are spaced evenly around the circle, they will stay evenly spaced forever. Although the oscillators run around the circle, their uniform spacing is unaltered. The nagging unsolved problem was whether this equilibrium is stable like the water in the glass, or unstable like the pencil balancing on its point. If it is unstable, it would mean that sync would emerge spontaneously, that the runners would eventually wind up in a pack.

That question had been festering for 15 years. Kuramoto himself had openly wondered about it. In his book he wrote that he could not see how to start. The question was bewildering, because there were infinitely many different ways for oscillators to be arranged incoherently. That was the rub. Incoherence was not a single state; it was a family of infinitely many states.

For years I couldn't see how to make any progress on the stability problem. Then, late one night, in the twilight before sleep, a strange image came to me: The oscillators aren't really like runners; they are like molecules in a fluid. Just as water is made of trillions of discrete molecules, this fictitious "oscillator fluid" would be made of trillions of discrete dots running around the circle.

The image was actually weirder than that. I needed to imagine a different fluid for each frequency in the distribution. Infinitely many different frequencies, like the blend of colors in the rainbow. So I pictured a rainbow of colored fluids, all swirling around the same circle, never mixing because oscillators never change their natural frequency. The advantage of this psychedelic formulation is that incoherence becomes a single state. Not an infinite family anymore, just one state of uniform density, with each colored fluid spread evenly around the circle.

I jumped out of bed and grabbed a pencil and paper. Dreamy ideas are often illusions, but this one felt right. The first step was to adapt the laws of fluid mechanics to my imaginary oscillator fluid. Then I wrote out the equations to set up the standard test for stability: disturb the system from equilibrium, solve the equations for the disturbances (these equations are solvable because they're linear, even if the original system is not), and check whether the disturbances grow or decay.

The equations showed that the answer depends on how similar the oscillators are. If they're identical, or nearly so, I found that the disturbances grow exponentially fast as oscillators clump together in phase, in an embryonic form of sync. Then out popped a formula for the exponential growth rate (analogous to the interest rate for how fast your money compounds in the bank). No one had ever found such a formula before. It was a definite prediction, either right or wrong. The next morning I would check it on the computer.

My hand was sweating as I wrote each new line of the calculation. It's all working. I'm seeing the birth of order. Then I paused. Is there a critical frequency spread at which the growth rate falls to zero, and incoherence is no longer unstable? Yes—the critical condition occurs at the same threshold that Kuramoto found. That was very reassuring. I had just found a new way to calculate the phase transition, the freezing point where spontaneous synchronization first occurs.

A few hours after the sun came up I called my collaborator Rennie Mirollo to fill him in. I started to describe my ideas about oscillator fluid, but it wasn't long before he interrupted. "What is this sophistry?" As a pure mathematician he had never studied fluid mechanics, and he liked his equations straight up, with no imagery attached. The whole calculation sounded fishy to him. But I was sure it was right. Later that day I went to the office and confirmed that the predicted growth rates were in perfect agreement with the results of computer simulations. Rennie quickly made his peace with oscillator fluid.

Together we worked out the stability of the incoherent state on the other side of the threshold, where the spread of frequencies is large, analogous to temperatures above the freezing point. We expected that incoherence should now become stable. But the equations were telling us instead that it was "neutrally stable"—a very rare, borderline case in which transient disturbances neither grow nor decay.

For example, picture a marble sitting at the bottom of a smooth hemispherical punch bowl. If you displace the marble from the bottom, it rolls back: The bottom is a point of stable equilibrium. Now suppose the bowl has an adjustable shape; by turning a knob, you can gradually morph it into a flatter shape, one with less curvature, like a giant contact lens. The bottom is still stable, but less so: A displaced marble rolls back more slowly. As you continue to turn the knob, the shape droops flatter and flatter, becoming dead level at a critical setting of the knob, and then droops so much that it becomes an upside-down contact lens, a gentle dome, finally becoming an upside-down hemisphere. During the morphing process, the bottom of the bowl has turned into the top of the dome. Now a displaced marble would roll down the side; the equilibrium has become unstable. The switch occurred at the critical boundary between stability and instability, when the contact lens was completely flattened. At that one setting of the knob, and only that one, the equilibrium is neither stable nor unstable. It's in a state of limbo: It's neutrally stable. A marble displaced from neutral equilibrium doesn't roll back, but it doesn't roll away either.

As this metaphor suggests, neutral stability normally occurs only at transitions, at critical settings of a system's parameters (the "knobs" that control its properties). But the Kuramoto model was breaking this rule. Its incoherent state was doggedly staying neutrally stable, even as we widened the bell curve to make the population more diverse. Turning that knob over a wide range of parameters made no difference.

We discussed this startling result with Paul Matthews, an instructor in applied math at MIT. Paul ran some computer simulations that only deepened the mystery. He tested the stability a different way, by computing the long-term behavior of the order parameter, and found that it decayed exponentially fast—normally the signature of stability, not neutral stability. Now we were truly mystified. Incoherence was neutral by one measure, yet stable by another.

A few weeks later, Paul gave a lecture at the University of Warwick in his home country of England, where he described all our strange results. One of the professors in the audience, George Rowlands, told Paul that what we were seeing was not so strange: It's called Landau damping, and plasma physicists have known about it for 45 years.

None of us knew much about plasmas, but we had all heard of Landau. Lev Landau was one of the supreme physicists of the twentieth century. In an era of specialization he had mastered every branch of theoretical physics, from subatomic particles to turbulence in fluids. He was a flamboyant, ornery genius whose career ended on January 7, 1962, when he was nearly killed in a car accident near Moscow. The crash shattered eleven bones, fractured his skull, punctured his chest, ruptured his bladder, and sent him into a coma. His brain waves went flat for 100 days, but his doctors maintained him on a respirator and would not let him die. On four occasions he was pronounced dead, only to be revived by heroic measures each time. Later that same year he was awarded the Nobel Prize for discoveries he had made a decade earlier, in which he used quantum theory to explain the weird behavior of superfluid helium at temperatures close to absolute zero. He was finally released from the hospital in October 1964; he never fully recovered and died a few years later.

Among his many contributions, in the late 1940s Landau had predicted a counterintuitive phenomenon about plasmas. Plasmas are sometimes called the fourth state of matter, farther up the temperature scale from solids, liquids, and gases. They're found in the sun and in thermonuclear fusion reactors, where ordinary atoms are boiled into an ionized gas made of roughly equal numbers of electrons and positively charged ions. The paradoxical phenomenon that now bears his name arises when electrostatic waves travel through a highly rarefied plasma. Landau showed that the waves could decay even if there were no collisions between the particles and no friction or dissipation of any kind. What George Rowlands had realized was that Landau damping is governed by essentially the same mathematical mechanism as the decay to incoherence in the Kuramoto model: The electrons in the plasma play the role of the oscillators, and the size of the ripples in the electric field they generate plays the role of the order parameter.

It seems amazing that there should be a link between the violent world of superhot plasmas in the sun, where atoms routinely have their electrons stripped off, and the peaceful world of biological oscillators, where fireflies pulse silently along a riverbank. The players are different, but their abstract patterns of interaction are the same. Once that link was exposed, we were able to transfer Landau's techniques to the Kuramoto model, answering a riddle that had lingered for years. There was also a payback from biology to physics. John David Crawford, a physicist at the University of Pittsburgh, was able to apply insights won from the study of biological synchrony to solve a long-standing problem about the behavior of plasmas.

The theories of how biological oscillators sync with one another have been successful from a mathematical perspective. They have shed light on one of nature's most fundamental mechanisms of self-organization. Still, a more tough-minded question is whether the models describe reality faithfully. Do they predict phenomena that agree with data from real fireflies, heart cells, or neurons?

We don't know. There have been no tests so far. The experiments are difficult because they require measurements at the level of individual animals or cells, especially their natural frequencies and their responses to stimuli of varying strength and timing; and at the level of the entire network, to quantify the interactions between oscillators and the resulting collective behavior. It's particularly hard to measure interactions between pairs of oscillators. If they are left in the network, the measurements may be confounded by the influences of other oscillators; if they are removed from the network, surgically or otherwise, the surrounding oscillators and connections among them may be damaged in the process. Furthermore, the connectivity of networks is typically unknown, except for a few small systems of neurons. Without knowing who is interacting with whom, it's impossible to test the models quantitatively. In a tree full of fireflies, for example, you would have to figure out exactly which bugs can see which, measure all their intrinsic flash rates one by one, and finally measure each insect's sensitivity and influence functions. No one has even tried this experiment for two fireflies, let alone a whole congregation of them.

A more qualitative test would be to confirm or refute the existence of a phase transition. The prediction is that the degree of synchronization should increase sharply, not gradually, as either the coupling strength or the frequency spread is tuned through a critical value. Here too, the experiment would be tricky. To change the coupling strength between fireflies, you could put them in a darkened room, and then adjust the ambient light level with a dimmer so that the insects would see each other more or less well. That's easy enough, but measuring all the simultaneous flash patterns of the insects would be taxing; without that information, there would be no way to determine the degree of synchronization, and hence whether a transition had occurred. The analogous experiment might be easier with neurons, but there you'd have to record from each cell simultaneously (again, technically very difficult) while administering drugs to uncouple them progressively, while taking care to ensure that the drugs don't change any other properties of the cells besides their mutual coupling. No one has tried it yet.

Or one could look for Wiener's spectrum of frequencies, with its narrow central peak arising from a dip on either side. That was the cornerstone piece of evidence for his theory of frequency pulling, and given its central role, it always seemed odd to me that I'd never heard about its being replicated. And something else was suspicious. If Wiener and his collaborators had really found the smoking gun—the double-dip spectrum that he believed to be the mark of synchronization—why didn't he let the data speak for itself? In his 1958 book,

Nonlinear Problems in Random Theory, he offered the schematic picture of the spectrum we have seen earlier, with its perfectly symmetrical peak rising from a perfectly symmetrical double-dip, all centered at exactly 10 cycles per second. No one could be fooled by that. His graph didn't even include tick marks on the axes. Then, in the 1961 edition of *Cybernetics,* he finally presented some real data (presumably the most convincing example he had), yet his beloved dip was nowhere in sight.

Some years ago, I asked Paul Rapp, a mathematical biologist and an expert on brain waves, if he had ever run across that spectrum in his own measurements. No, he had not, but he thought it should be easy to find, if it were real. He conducted a series of new experiments, specifically looking for the effect, and even with today's improved technology, he couldn't find it. So was Wiener deluding himself? Was the dip nothing more than a figment of his fertile imagination? I didn't want to believe that, so it came as a relief to learn the inside story about what really happened back in 1958.

While attending an applied math conference in July 2001, I happened to strike up a conversation with Jack Cowan, a theoretical biologist who has long worked on mathematical models of the brain. Given the likelihood that he would know a great deal about the alpha rhythm, I asked whether he was familiar with Wiener's old theory. Oh yes, he said with a smile—he had been at MIT as a postdoctoral fellow at the time. Wiener had buttonholed him and lectured him "two hundred times" about that peculiar spectrum. "Norbert loved to capture people to provide an audience for himself."

Cowan arrived at MIT in the fall of 1958 to work as a postdoctoral fellow in the communications biophysics group led by Walter Rosenblith. Around that time, Margaret Z. Freeman, a research associate in Rosenblith's group, made the first measurements of the spectrum, and it was she who discovered the signature peak and double dip, much to Wiener's delight. Though the results were still in preliminary form, Wiener happily crowed about them in his 1958 book.

Unfortunately, Freeman's results were wrong. "Other people tried to replicate her findings," Cowan told me, "and when they couldn't, the whole thing sort of fizzled out." Freeman had made an error in her computations. When she checked them again, the dip disappeared.

It was too late for a retraction. Wiener had already published his book showing the schematic drawing of the spectrum. But three years later, in *Cybernetics,* he would have a chance to correct the error. This time, he chose to show real data. Here's how he describes the spectrum:

> When we inspect the curve, we find a remarkable drop in power in the neighborhood of frequency 9.05 cycles per second. The point at which the spectrum substantially fades out is very sharp and gives an objective quantity which can be verified with much greater accuracy than any quantity so far occurring in electroencephalography.

That's the sound of Wiener at his confident best, the ex-prodigy teaching the encephalographers a thing or two. But then his language turns tentative, his mood subjunctive:

> There is a certain amount of indication that in other curves which we have obtained, but which are of somewhat questionable reliability in their details, this sudden fall-off of power is followed quite shortly by a sudden rise, so that between them we have a dip in the curve. Whether this be the case or not, there is a strong suggestion that the power in the peak corresponds to a pulling of the power away from the region where the curve is low.

When I first read this ten years ago, I was struck by the crabbed language. It's not like him—normally Wiener's writing is bold and direct. But when I read it now, the passage seems almost poignant. I think I can hear the sound of a man struggling with himself, a scientist clinging to an idea that he knows must be right, while summoning the strength to be intellectually honest. Although the dip is nowhere to be found, he asks us to believe that it occurs in other records, but he won't allow himself to push too hard; he admits that those other records are "questionable" and says that there's only a "certain amount of indication" of a dip in them. And whether the dip was there or not, the last sentence shows that he was not going to give up on the idea that oscillators synchronize by pulling on one another's frequencies. He felt sure that it was a universal mechanism for sync. It was bound to be important. He refused to fall victim to what T. H. Huxley called "the great tragedy of science—the slaying of a beautiful theory by an ugly fact."

Wiener was like a prophet, with a vision of how the world should work. We see that tendency in other great scientists. Galileo would not have discovered that a body in motion tends to stay in motion (the law of inertia) if he had been content to describe what really happens (friction causes things to stop). By disregarding the inessential, he discovered the most fundamental law of mechanics. Gregor Mendel discovered the laws of genetics by studying the inheritance patterns of peas. Some modern statisticians have questioned his data, calling it too clean to be credible, while others suggest more generously that Mendel carefully chose the peas that would best illustrate the principles he sought to propound. Whichever version you believe, it seems clear that Mendel knew exactly what he was looking for.

Although Wiener was wrong about the alpha rhythm, the irony is that he was right about a different kind of rhythm in the brain. In 1995, the biologists David Welsh and Steve Reppert at Massachusetts General Hospital discovered that the brain does contain a population of oscillators with distributed natural frequencies, which do pull one another into synchrony, and which do make a more accurate oscillator en masse than individually. Wiener anticipated all that, but he missed an important detail: Instead of cycling 10 times per second, these cells cycle about a million times slower. These are the cells of the circadian pacemaker, the internal chronometer that keeps us in sync with the world around us.

· *Three* ·

SLEEP AND THE DAILY STRUGGLE FOR SYNC

———————

Like all newborns, my daughter Leah was an anarchist for the first three months of her life. She fed and slept on no discernible schedule. By the time she was 11 months old, she slept through the night, though with one small problem: She'd invariably wake my wife and me up at 5:20 a.m. She'd grab the rails of her crib, hoist herself up to a standing position, and tactfully cough a few times to signal that she was ready for her morning bottle. We knew we shouldn't complain (many parents endure far worse), yet we still wished that she'd get up at least an hour later. To coax her in that direction, we tried keeping her up late one night. Naturally this strategy backfired: The same little coughs wafted out of her bedroom at precisely 5:20 the next morning, but because she had slept less, she punished us with crankiness for the rest of the day.

Both of these timing problems were fundamentally failures of synchronization. As a newborn, Leah couldn't synchronize at all; her sleep-wake and feeding rhythms (to the extent she had any) wandered erratically relative to the world's daily cycles. At 11 months she presented the opposite problem: Now her rhythms were all too stubbornly synchronized, welded to a particular 24-hour pattern that we happened to find oppressive.

And it's not only babies and their parents who suffer from disrupted sync and its attendant sleep disorders. American society is gradually coming to realize that teenagers love to stay up late and have trouble getting up for school in the morning, not because of their sluggish natures or moral turpitude but because their internal body clocks are set differently, somewhere in a time zone to the west of us. At the other end of the spectrum, many elderly people wake up in the early morning while it's still pitch-black outside, and then can't fall back asleep, tired as they may be.

Other kinds of sync disorders have nothing to do with age. We bring some of these on ourselves, with our round-the-clock work schedules. Think of the medical and family problems that plague tens of millions of nurses, truck drivers, nuclear power plant operators, and other workers who rotate between day and night shifts; the industrial accidents at Bhopal, Chernobyl, and Three Mile Island, all of which occurred between midnight and 4 A.M.; the fuzzy-headedness and errors in judgment caused by jet lag: These too are the by-products of deranged sync, of mismatches between our bodies and the demands of the new 24-hour society.

When you start to think about it, it's miraculous how easily we stay in step with the world. Blind people, however, don't take this for granted: Most of them are unable to keep to a 24-hour schedule. They roll in and out of phase with the rest of society every few weeks, which can make it difficult to maintain jobs and social obligations. As one blind woman put it, "Being blind is okay, although something of an inconvenience. Having a free-running sleep cycle can be awful."

So the rest of us should cherish the miracle of sync. Of course, we rarely give it a thought, since it occurs spontaneously. Millions of years of evolution have tuned our bodies to harmonize automatically with the cycle of day and night. But how does this work? We speak about body clocks, but are they real or just figures of speech? Where are they located: In our brains or in every cell? What is their biochemical mechanism? How do they synchronize one another, and what aligns them to the cycle of day and night? After decades of research, much of it slow and frustrating, the answers to some of these mysteries are finally within reach. The study of biological clocks has become one of the hottest fields in science today.

The picture that is emerging suggests that we are like wheels within wheels, hierarchies of living oscillators. Or to put it more vividly, the human body is like an enormous orchestra. The musicians are individual cells, all born with a sense of 24-hour rhythm. The players are grouped into various sections. Instead of strings and woodwinds, we have kidneys and livers, each composed of thousands of cellular oscillators, similar within an organ, different across organs, all keeping a 24-hour biochemical beat but entering and exiting at just the right times. Within each organ, suites of genes are active or idle at different times of day, ensuring that the organ's characteristic proteins are manufactured on schedule. The conductor for this symphony is the circadian pacemaker, a neural cluster of thousands of clock cells in the brain, themselves synchronized into a coherent unit.

Sync enters at three different levels. At the lowest, most microscopic level, the cells within a particular organ are mutually synchronized; their chemical and electrical rhythms vary in lockstep. At the next level, synchrony occurs between the various organs, in the sense that they all keep to the same period, even though the cells have differentiated into disparate types. This kind of sync occurs within the body itself, and so is called internal synchronization. It doesn't mean that all the organs are active at the same times. On the contrary, some are silent while others are going strong. The sync is in the sense of period matching, keeping the same beat, just as musicians keep the beat in their heads even when they are quietly awaiting their turn to play. Finally, the third level of synchrony is that between our bodies and the world around us. Under normal conditions of living on a regular schedule, seeing sunlight, sleeping at night, and so on, the entire body is synchronized to the 24-hour day, driven mainly by the cycle of light and darkness. This process of external synchronization, of falling in step with the outside world, is called entrainment.

At the moment, our best theories of human circadian rhythms are more descriptive than mathematical. That is by necessity—we lack a deep under-standing of the system's architecture and dynamics. Its hierarchical organization appears to be far more complex than anything envisaged by the simple models of oscillator populations discussed earlier. A congregation of fireflies could be approximated as a collection of self-sustained oscillators, all of which are identical or nearly so, all firing at about the same time. In that sense, the level of complexity associated with synchronous fireflies is comparable to that of a single organ, or in musical terms, a single section of the orchestra. We are just beginning to learn how the sections play together as an ensemble, and how the pacemaker orchestrates them all. In other words, we are trying to learn the rules of the circadian symphony.

We know that such rules exist, because we can see their manifestations at a larger scale, in the behavior of a whole, integrated human body: in our daily rhythms of sleep and wakefulness, hormone fluctuations, digestion, alertness, dexterity, and cognitive performance. At this higher level, scientists have recently discovered cryptic regularities in the timing of human sleep-wake cycles and other circadian rhythms, even though the microscopic basis for these laws remains enigmatic. In that respect, our present situation parallels the early development of genetics. Mendel discovered that various characteristics of pea plants were passed on to their offspring according to certain mathematical laws, and realized that these patterns could be explained by postulating hypothetical entities called genes that recombined according to certain rules. All this was done long before any knowledge about the reality of genes and their physical embodiment in strings of DNA. Similarly, we now know that human circadian rhythms obey their own kinds of laws, though we remain in the dark about their fundamental biochemical basis.

With respect to sync's impact on our everyday lives, one of the most imme-diate issues is how the circadian pacemaker affects sleep. That part of the puz-zle has been largely worked out, thanks to dramatic experiments in which brave volunteers lived alone for months in underground caves or clockless, window-less apartments, isolated from all knowledge of the time of day, free to sleep and wake whenever they felt like it. The results of those studies turned out to be so bizarre, yet laced with such tantalizing hints of pattern, that Art Winfree was led to proclaim, "A Rosetta stone has appeared in our midst." By deciphering the circadian code, scientists and doctors are learning how to design better shift-work schedules and treat some forms of insomnia that were previously intractable. They have even explained some of life's little mysteries—like why so many cultures take an afternoon siesta, or why we often have trouble falling asleep on Sunday nights.

On February 14, 1972, Michel Siffre gazed across the arid landscape near Del Rio, Texas, and savored the last sunlight he would see for six months. Then he smiled bravely for the television cameras, hugged his mother and kissed his wife good-bye before descending down the 100-foot vertical shaft into the bowels of Midnight Cave. Awaiting him underground was a campsite stocked with scientific equipment, furniture, a nylon tent, freezers, food, and 780 one-gallon jugs of water.

Siffre, a French geologist and sleep researcher, was about to be his own guinea pig in the most elaborate time-isolation experiment ever performed. Assisted by NASA, he and his research team wanted to study the basic rhythms of human life in the absence of clocks, calendars, and all other daily time cues. He'd tried this once before. Ten years earlier, in the first such experiment ever conducted on a human subject, he endured two lonely months in the numbing cold of an underground cave in the Alps, only to emerge, as he put it, a "half-crazed, disjointed marionette." That ordeal had provided the first scientific evidence that human beings have innate circadian clocks, with a cycle length slightly longer than 24 hours.

Now, in the constant balminess of Midnight Cave, where the temperature was always 70 degrees Fahrenheit, Siffre hoped for a more pleasant experience. If anything, it was worse this time. His mind nearly collapsed from the strain of six months alone in a cave. His record player broke and his books became too mildewed to handle. To relieve the boredom, he tried to capture his one companion, a tiny mouse, by enticing it with some jam, only to kill it inadvertently when the makeshift cage—a casserole dish—struck its head. The months of lethargy and bitterness wore on. On day 79, Siffre phoned his collaborators on the surface, begging to be released—*"J'en ai marre!"* (I've had enough!) Yes,

yes, all is well, everything is going fine, they told him. Squinting in the darkness, breathing cave dust mixed with bat guano, he began to consider suicide. On the final day of the experiment, nature almost accommodated him: He received a blast of electricity through the electrodes recording his heart rhythm, perhaps when a distant lightning bolt struck the surface of the earth and leaked into the wires. It was a measure of how far his wits had deteriorated that it took three more shocks before he thought to disconnect the equipment.

Fortunately, the experiment produced some remarkable results. During his first five weeks in the cave, Siffre unknowingly lived on a 26-hour cycle. He woke up about two hours later each day and drifted around the clock relative to the outside world, but otherwise maintained a normal schedule, sleeping about a third of the time.

Meanwhile his body temperature waxed and waned, just as it normally does in everyone, every day. This may come as a surprise: Contrary to what many of us have been taught, body temperature in a healthy person does not stay constant at 98.6 degrees Fahrenheit, or at any other number; it typically undulates through a range of about 1.5 degrees over the course of a day, even if we lie in bed and don't exert ourselves. As the physician William Ogle first reported in 1866, "There is a rise in the early mornings while we are still asleep, and a fall in the evening while we are still awake. . . . They are not due to variations in light; they are probably produced by periodic variations in the activity of the organic functions."

Now Siffre was confirming what Ogle had so presciently guessed a century earlier about the origin of the body temperature cycle. In the constant conditions of his cave, Siffre was oblivious to day and night, and had no other clock to go on besides the internal rhythms of his own physiology. Divorced from the influence of the 24-hour world outside, his "organic functions"—as reflected by his body temperature—oscillated in sync with his sleep-wake cycle at the same idiosyncratic 26-hour period. In fact, he always went to bed when his temperature bottomed out, although he was unaware of this.

At this stage in the experiment Siffre was behaving like a hamster or a fruit fly or any other organism that has ever been studied in time isolation.

Some creatures cycle a bit faster than 24 hours, some a bit slower—hence the term *circadian rhythm,* from the Latin: *circa* means "about," and *dies* means "a day." For example, a laboratory mouse confined to a cage and kept in constant darkness will happily jump on its running wheel at a predictable time, about half an hour earlier than the day before, and then run for miles. Thus the mouse has a circadian rhythm of activity with an intrinsic period of 23.5 hours. A mimosa plant kept in constant artificial light will open and close its leaves on a cycle of 22 hours. Virtually all living things, from monkeys to microbes, show similarly persistent rhythms when allowed to "free-run" in the absence of time cues.

On day 37 of his experiment, however, Siffre lost his resemblance to all other species. His body did something strange, something uniquely human: His sleep and body temperature rhythms came unglued. He stayed up way past the nadir of his body temperature cycle, essentially pulling an all-nighter, after which he slept for 15 hours, double his usual amount. For the next month, he bounced back and forth on a wild schedule, sometimes keeping to his original 26-hour pattern, only to follow it, unaccountably, with yet another whopping sleep-wake cycle, 40 or 50 hours long. Yet Siffre perceived none of this. And through it all, his temperature rhythm never budged from its 26-hour pace.

This weird phenomenon is called spontaneous internal desynchronization. Its implication is that two circadian rhythms (sleep and body temperature) can run at different periods in the same organism. Ever since it was first reported by the German biologist Jürgen Aschoff in 1965, researchers have been perplexed by this sudden breakdown of the body's temporal order, all the more so since plants and animals never desynchronize internally. When Siffre examined his own data, he too was mystified. "Jagged, seemingly random," was how he described it three years later.

We now know that Siffre's sleep-wake cycle was not random. In fact, it obeyed beautifully simple mathematical rules. What's even more astonishing, the same rules have been found to hold for all human subjects who have ever been studied in time isolation. The first hints of this universal structure were uncovered by a young graduate student working at a hospital in New York, a newcomer who would go on to become the world's leading authority on human circadian rhythms.

In the mid-1970s, Elliot Weitzman and his student Charles Czeisler decided to try their hand at time-isolation experiments. There were only three other groups in the world working in this field: Siffre's in France, Aschoff's in Germany, and one led by John Mills in England. It was an expensive, elaborate undertaking, to say the least, but the potential payoffs to human medicine and biology were compelling.

On the fifth floor of one of the old wings at Montefiore Hospital in the Bronx, Weitzman and Czeisler built a soundproofed, windowless facility consisting of three one-bedroom suites and a control room in the middle. They placed ads in the paper to recruit candidate subjects, hoping to attract craftsmen, artists, or graduate students with a thesis to finish: in short, anyone with a long-term project or some other good reason to get away from the world for one to six months. The subjects had to be screened psychologically. It would be disastrous if someone freaked out and had to quit the experiment in the middle, since the studies cost about $1,000 a day.

In return, these subjects enjoyed a life of pampered leisure. They were paid a few hundred dollars a week, given room and board, and allowed to live as they wished. They could wake up and sleep whenever they pleased. They could read, work, exercise, or listen to music, and ask for meals to be brought to them. They could even read newspapers or magazines, provided that the reading matter was long out of date. On the other hand, they could not wear watches, make phone calls, or listen to the radio or watch television, since those could be used to determine what time it was. The point of the time-isolation protocol was to observe human circadian rhythms in their most pristine form, uncorrupted by the influences of the outside world. For the same reason, the subjects were forbidden to drink coffee, tea, or alcohol, or to take sleeping pills, stimulants, or recreational drugs, all of which can disrupt the normal rhythm of sleep and wakefulness. (Earlier studies on animals indicated that alcohol and caffeine might even reset the circadian clock itself, though this effect seems to be minor compared with the familiar sedative or stimulant action of these chemical agents.)

Day after day, week after week, Weitzman and Czeisler monitored the subjects' alertness when they were awake, brain waves when they were asleep, and body temperature and hormone levels around the clock. For instance, to track the rapidly fluctuating profiles of growth hormone and cortisol (the body's stress hormone), they stuck an indwelling catheter in the subject's arm for the duration of the experiment, so that lab technicians could withdraw tiny blood samples every 20 minutes. Meanwhile, a rectal probe (like a piece of string) measured the subject's core body temperature continuously. To prevent spurious blips in the temperature record, subjects were told to remove the probe in the shower or if they needed to masturbate.

Unlike Siffre in the cave, they were not socially isolated and did not suffer any psychological injuries. They could chat with the technicians and often befriended them. Of course, the staff had to be careful not to disclose anything about the time of day. For example, the male doctors and technicians always shaved before entering the apartment so that their five o'clock shadows would not be a tip-off. All staff members greeted the subject by saying hello, not good morning or good evening, and a computer assigned them to work at random hours, so the subject couldn't tell what time it was by who was on duty. (Given their crazy schedules, it might have been equally interesting to study the staff's circadian rhythms.)

One of the former subjects recalled what the experience was like:

> When I was out of college I was broke and this was a way of making some money. . . . I spent a lot of time reading and writing to make up some incompletes. I got more done in a month than I normally did in a whole semester. I thought it was important to have a certain routine to maintain a measure of sanity, so I wore a shirt and tie, and shaved every day. One of my biggest problems was that my pants were wool and I couldn't get the creases pressed. So sometimes I walked around with a shirt and tie and shorts!
>
> Sometimes I felt like a prisoner, trading my youth for money. Although I didn't feel crazy, I thought others might think I was. I'm quite comfortable with

myself a little confined. I was happy as a clam. I could tell they were also a little strange, more interested in my urine samples than in some fascinating dreams.

They took blood samples every fifteen minutes. I had a catheter in my arm, and a butt probe, and all these things were attached to a movable pole. The first few days there was a definite presence but after the first week it became a part of you. It was like having a tail. . . .

I never knew what time it was and I didn't worry about it, except one time a technician came in with tuna fish on his breath and bloodshot eyes. I said, "Pretty tough night, eh?"

· · ·

Of the first 12 subjects that Czeisler and Weitzman studied, 6 of them internally desynchronized. For whatever reason, these subjects repeatedly stayed awake and asleep for extraordinarily long times, just as Siffre had done in Midnight Cave. A few maintained that odd schedule indefinitely, resulting in sleep-wake cycles that were 40 hours long, on average. Others regularly alternated between long cycles and more conventional ones, while still others would systematically lengthen their cycles as the experiment progressed, until they were sleeping only once every two days, without realizing it. There seemed to be no rhyme or reason to any of this.

Czeisler was especially intrigued by the long sleep episodes. Why would someone sleep for 15 hours straight? Could it be explained by how long the subject was awake beforehand? That would make sense: After staying up late, the subject might need to sleep more. But when Czeisler graphed the duration of sleep against the duration of prior wakefulness, he saw nothing. The graph was a blob. Although a statistical test for correlation showed a weak tendency for long sleeps to follow long wakes, it was unconvincing. By eyeballing the data, he could spot plenty of counterexamples where longer wakes were followed by *shorter* sleeps.

Meanwhile, the round-the-clock physiological measurements showed that the subjects' rhythms of body temperature, cortisol secretion, and alertness always remained rock-solid, running with a period just a little bit longer than

24 hours. No matter how erratic the sleep-wake cycle became, these three internal rhythms were always remarkably stable. More than that, they always moved in lockstep: Their periods were identical. That had to be a clue.

Czeisler tried another approach. He graphed the sleep and body temperature cycles together in a two-dimensional format called a raster plot. Circadian biologists had been making this type of plot for decades. It was the standard way to depict the leaf-opening rhythm of a plant or the wheel-running rhythm of a mouse, but it hadn't been used much for humans. The term comes from an analogy with television technology, where a process called rastering converts a continuous torrent of electronic information into a two-dimensional picture. Similarly, a raster plot takes the stream of data coming from an experiment and converts it into a two-dimensional graph. The raster chops the data into 24-hour blocks, and then stacks them vertically like a pile of bricks.

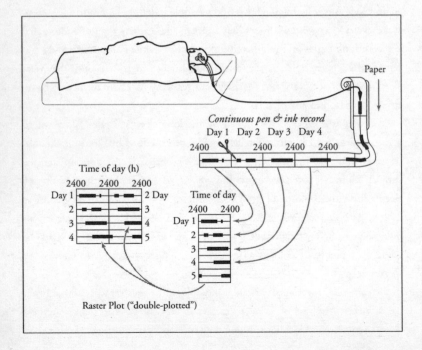

Day 1 is on the top of the pile, with day 2 directly below it, and so on, continuing until the last day of the experiment at the bottom of the pile. To summarize the subject's circadian rhythms on a given day, a black bar shows the hours when he was asleep, and a gray bar shows when his body temperature dipped below its average value. The virtue of raster plots is that any repetitive patterns in the data jump out at you. A strict 24-hour rhythm is instantly recognizable as a vertical stripe of bars, all starting and ending at the same time of day. A rhythm longer than 24 hours is a diagonal stripe that slopes down to the right.

When Czeisler made a raster plot for one of his desynchronized subjects, he immediately noticed that all the long sleep episodes—the mysterious ones—lined up diagonally. So did the short sleeps, but on a different line. And both lines ran parallel to a diagonal stripe formed by the trough of the body temperature rhythm.

The implication was startling. Even though the sleep-wake cycle had osten-

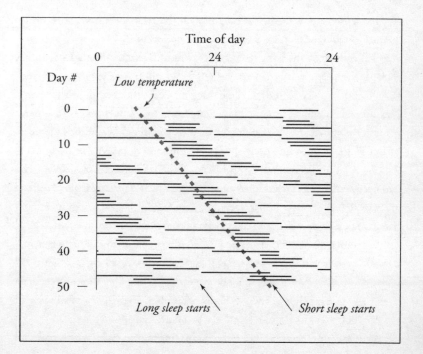

sibly come unglued from the temperature cycle, there was an ongoing, consistent relationship between them: Long sleeps always began at high temperature and short sleeps at low temperature. Czeisler checked the records for his other subjects, and the same rule worked every time. He reanalyzed old data published by the groups in France, Germany, and England. The rule had been hiding in there all along.

Czeisler had cracked the circadian code. By studying sleep in relation to the cycle of body temperature (rather than in relation to the time of day or any other external variable), he had discovered a natural reference frame, a natural measure of what time it is in the body. When viewed from this perspective, data that had previously appeared "jagged" and "random" suddenly lined up and snapped into place. How long a subject stayed asleep did not depend on how long he had been awake beforehand; it depended on when he fell asleep in relation to his cycle of body temperature.

To flesh out the mathematical form of the relationship, Czeisler made another graph, now plotting the duration of dozens of different sleep episodes versus the phase of the body temperature cycle at bedtime. In other words, he took all the sleeps that began when body temperature was low, and grouped them together. Then he did the same thing for sleeps that began near the temperature maximum, and so on. This allowed him to compare apples to apples; his raster plot had already shown him that sleep episodes beginning at similar phases in the temperature cycle should be similar in duration. He pooled the data from all his desynchronized subjects—some young, some old, some who had lived on 30-hour cycles, some on 40. Despite their drastic individual differences in all other respects, their sleep durations fell neatly into a single cloud of data points, a slightly blurred version of a universal mathematical curve.

Whenever the subjects happened to go to bed near the peak of their temperature cycle, the subsequent sleep episode was always very long, averaging 15 hours. Conversely, when they fell asleep near the time of minimum temperature, they slept much less, about 8 hours on average. Viewed across all phases, the cloud of sleep durations resembled a sawtooth-shaped wave.

That was unexpected. Unlike the body temperature cycle, which had the

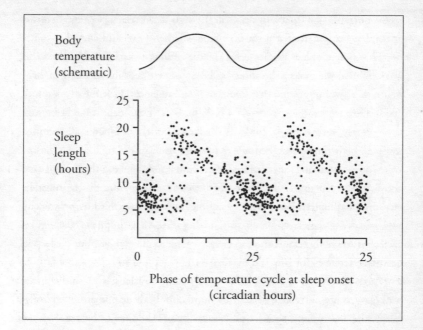

more familiar appearance of a sine wave, the cloud was strikingly asymmetrical. Sleep duration jumped vertically from 7 hours to 18 hours for episodes beginning about 9–10 hours after the temperature minimum, followed by a gradual, ramping descent back down to shorter sleeps.

The ramp implies something counterintuitive. When subjects go to sleep later in their body temperature cycles, they actually sleep *less,* even though they have been awake longer. The same peculiar pattern has been observed in the irregular sleep of train drivers and other night-shift workers, and you may have noticed it in your own sleep after a late-night party. When you finally get to bed, hoping for a long recovery sleep, you're exasperated to find yourself waking up after only 5 or 6 hours of tossing and turning. The problem is that the body's internal alarm clock is ringing. Time-isolated subjects nearly always awaken in the first few hours after their temperature starts rising, at around the same time that cortisol (the body's stress hormone) is being secreted maximally and alertness is revving up. The same is true during 24-hour entrainment. So if we go to sleep later, we tend to wake up sooner, because the alarm starts ringing whether we've slept enough or not.

That's the rough explanation for the descending ramp. To understand the vertical jump, where sleep duration can be either very short or very long or anywhere in between, imagine that you stay up all night until the following afternoon. Then, if you allow yourself a little mid-afternoon nap, it might be just that—a nap—or you might conk off for the rest of the afternoon, all evening, and sleep clear through to the following morning.

This explanation relies on an implicit conversion between the time of day and "circadian phase" (the phase of the body temperature cycle, the only measure of time that has physiological meaning in conditions of protracted time-isolation). To extrapolate to the real world, where both the sleep-wake cycle and the body temperature cycle are entrained to the 24-hour clock, we need to reference all biological events to the low point in body temperature. The proper conversion formula emerged from Czeisler's later studies: For people who are entrained to the 24-hour day, body temperature typically reaches its lowest point about 1 or 2 hours before the time of habitual wake-up. For example, much of the labor force wakes up at around 6 or 7 A.M. Hence, for those people, minimum body temperature probably occurs between 4 and 6 A.M. The jump in sleep duration is predicted to occur about 9–10 hours after that, which translates to a clock time of 1–4 P.M. As claimed, that's nap time.

Czeisler and Weitzman found that many other physiological and cognitive rhythms were linked to the phase of the body temperature cycle. For instance, they asked their subjects to assess their alertness at frequent times while they were awake. The subject was handed a non-numeric, continuous vertical scale marked only "very alert" at the top, and "very sleepy" at the bottom, and asked to draw a line at the appropriate level. (Numbers were omitted on the scale to discourage the subject from automatically repeating his previous assessments.) The results showed that alertness goes hand in hand with body temperature: It's low when temperature is low and high when temperature is high.

Again using the conversion formula above to translate circadian phase to clock time, these time-isolation data predict that during 24-hour entrainment, minimum alertness should occur around the time of the temperature trough, namely, 4 to 6 A.M. That's a notorious time of day. The accident at the Three Mile Island nuclear-power plant occurred then, with a crew that had been on night duty for just a few days. Chernobyl, Bhopal, *Exxon Valdez:* All those disasters occurred in the middle of the night, and were tied to human error. Field studies show that from 3 to 5 A.M., workers are slowest to answer a telephone, slowest to respond to a warning signal, and most apt to read a meter wrong. It's a bad time to be awake, especially if you are required to do something monotonous and important. Shift workers call it the zombie zone.

Even if you've never worked the night shift, you've probably noticed your alertness rhythm during an all-nighter. The later you stay up, the groggier you become. At some point, usually between 3 and 6 A.M., your eyes start to itch. The desire to sleep becomes overwhelming. After even more sleep deprivation, out of nowhere comes a second wind and you start to feel better. You've just gone through the trough of your circadian cycle. Now alertness starts to rise, along with temperature and cortisol secretion. The interesting point is that this same sleepy time shows up in the time-isolation data, even though the subjects are well rested and are not working the night shift. The zombie zone is built into our biology.

Along with modulating alertness and the duration of sleep, the circadian clock also regulates the internal structure of sleep, specifically the propensity for rapid-eye movement (REM) sleep. REM is a bizarre state, perhaps more so than most people realize. We dream vividly, and our eyes dart from side to side. Our breathing and heart rate fluctuate erratically. Spinal inhibition paralyzes the body—a good thing, since it prevents us from acting out our dreams. (In experiments on cats where the spinal inhibition was blocked, they ran around while still in REM, as if chasing imaginary mice.) Men normally have erections during REM sleep. That involuntary tumescence enables doctors to distinguish psychological from physical impotence; they wrap a roll of postage stamps around the penis at bedtime, and if the patient wakes up with torn stamps, the problem is psychological.

REM is as different from non-REM sleep as it is from wakefulness. Sleep used to be viewed as a bland, uniform state, with the body and brain shut down for the night. Going to bed was like putting the car in the garage and turning off the engine. Now we know that the engine—the brain—is never turned off. In non-REM sleep, the car (the body) is in the garage with the engine running, but set in neutral so the car won't move. In REM sleep, the car is in the garage with the engine running, and both the gas pedal and the brake are floored. (The gas pedal represents the brain revving furiously; the brake is the spinal inhibition that keeps the body from moving.)

REM sleep has its own rhythm of occurrence, much faster than the circadian rhythm. The brain cycles through various stages of sleep about every 90 minutes. After we crawl into bed, we first slip from wake into light sleep; then into deep sleep, where the brain waves are large and slow; and then back out to light sleep and into REM for the first of several dreams. The first dream period is usually short, about 10 or 20 minutes. The REM episodes generally lengthen as the night goes on, so that by the early morning hours, we may be treated to a full hour of surreal entertainment, or possibly a horror movie.

For people who are normally entrained to the 24-hour day, the peak time for REM is in the early morning, near the end of sleep. That explains why we so often wake up after a long dream, and why men so often wake up erect. But this commonplace association of REM with the end of sleep is actually the wrong generalization. That is not the law of REM. The correct law was discovered by Czeisler and Weitzman in their time-isolation experiments. When they initially measured the brain waves of their subjects, they were shocked to find that REM accumulated most rapidly near the beginning of sleep, not near the end. Moreover, that's when the longest REM episodes occurred. Both results seemed topsy-turvy, counter to everything taught in medical school. In fact, REM at the onset of sleep is normally very rare and diagnostic of narcolepsy, a debilitating sleep disorder.

The paradoxical results began to make sense when Czeisler and Weitzman realized what the true law of REM is. The propensity for REM is synchronized to the body temperature cycle, not to sleep itself. The brain doesn't care whether it is the end of sleep or the beginning—what matters is what time it is in the body. The rule is that REM is most likely just after the part of the temperature cycle when your body is coldest. In 24-hour entrainment, that's the circadian phase when most people wake up, which is why REM is so common at the end of sleep. In contrast, free-running subjects often *fall asleep* around the temperature minimum, which is why they often have REM at sleep onset. There's nothing pathological about it.

The circadian rhythms of sleep duration, alertness, and REM propensity are not the only ones to march in lockstep with the body temperature cycle. Later studies demonstrated that our rhythms of short-term memory, the secretion of the brain hormone melatonin, and several other cognitive and physiological functions also run at the same period and maintain constant phase relationships to the temperature cycle and to one another. There is only one simple way to explain how all these disparate rhythms could be so tightly linked: They must all be controlled by the same biological clock.

For many years, this circadian pacemaker was nothing more than a hypothetical entity. Its existence was inferred indirectly, just as atoms were in the nineteenth century. The search for its location in the body always had the potential to degenerate into an endless chase based on a wrongheaded question. After all, early experiments on single-celled algae had shown that even they could exhibit circadian rhythms. So for more complex, multicellular creatures like ourselves, it might be that the whole organism is made of trillions of clocks. In other words, we might not have a clock; we might *be* a clock.

And that spooky thought is turning out to be right. For 30 years we've known that adrenal glands and liver cells can display circadian rhythms of their own, even when they are removed from the body and kept alive in a dish. The same now appears to be true of heart cells and kidney cells. Clock genes are turning up in tissues everywhere in the bodies of fruit flies and small mammals like mice and hamsters; presumably we, too, are congregations of circadian oscillators.

Still, there has always been strong reason to believe that, in mammals at least, all these peripheral clocks are ruled by a single master, probably localized somewhere in the part of the brain called the hypothalamus. As far back as the early 1900s, doctors had noticed that patients with tumors in this area suffered from irregular sleep-wake cycles. An even more telling piece of evidence came from the work of Curt Richter, a biologist at Johns Hopkins University, who spent almost 60 years stalking the circadian pacemaker. In an arduous and gruesome series of experiments, Richter blinded rats and then systematically removed their adrenals, pituitaries, thyroids, or gonads; induced convulsions; and administered electroshock, alcoholic stupor, and prolonged anesthesia. After sewing the rats back up and returning them to their cages, he found that none of these horrific interventions altered their free-running activity rhythms. The clock was still ticking. Then he cut their brains in one location after another, testing whether any individual lesion disrupted their circadian rhythms. None of the nicks made any difference. The rats went right on feeding, drinking, and running rhythmically—except when the lesions were placed in the front part of the hypothalamus. Then the rats became arrhythmic.

In the 1970s, other researchers pinpointed the clock even more precisely. Guided by the fact that cycles of light and dark could entrain circadian rhythms, they injected the eyes of rats with radioactively labeled amino acids, hoping to trace the neural pathways from the retina back to the putative clock. Along with the expected pathways to the brain's visual centers, they also found a monosynaptic pathway—a neural hotline—to the suprachiasmatic nuclei, two tiny clusters of neurons in the front of the hypothalamus. This neural architecture was extremely suggestive. Apparently the clock was so important to an animal's survival that evolution had joined it to the eyes by a dedicated line, rather than pausing to make several synaptic hops. To settle the matter once and for all, the researchers then surgically destroyed the suprachiasmatic nuclei and found that the rat's circadian rhythms disappeared along with it. The master clock had finally been found.

The details of how the pacemaker works are still sketchy. It's known that many of the thousands of neurons in the suprachiasmatic nuclei are oscillators. They spontaneously cycle through a cadence of electrical firing each day, driven by the waxing and waning concentrations of molecules called clock proteins. These molecular circadian rhythms are themselves generated by an interlocking set of biochemical feedback loops, involving DNA transcription and translation of something like eight clock genes (at last count—this research is in constant flux). Then, somehow, thousands of these oscillating "clock cells" manage to synchronize their electrical activity, coupled perhaps by chemical diffusion of a neurotransmitter called GABA. Finally, the collective electrical rhythm of the pacemaker is conveyed—again, through unknown means—to the peripheral oscillators in the liver, kidney, and other organs throughout the body, disciplining them to run at the same period as the master clock.

The explanation of Czeisler's results, then, is that all the rhythms he was measuring were coordinated by a single circadian pacemaker. The body temperature cycle is a reliable marker for it; that's why all the other rhythms lined up when viewed in that natural reference frame. We still have no idea how the pacemaker biochemically determines the duration of sleep or the propensity for REM. All that must wait for another day.

For now, we can only be awed by the performance of this brilliant maestro, mysteriously orchestrating dozens of rhythms inside us. When everything is working right—when we're not jet-lagged or otherwise desynchronized—the performance of the pacemaker is breathtaking. Consider how it steers the body through the most biologically stressful moment of every day: the moment of awakening. On the pacemaker's command, body temperature has already begun rising two hours earlier. The adrenal gland secretes a burst of cortisol to rouse us for the battles ahead. The internal alarm clock starts ringing. Rhythms of cognitive function, memory, dexterity—all turn on and begin to climb. As we go through the rest of the day, virtually every organ system and physiological function ebbs and flows on a predictable schedule. The silent symphony inside us explains why cancer chemotherapy is most effective at certain hours (reflecting rhythms in DNA synthesis and other cellular processes), and why heart attacks are most probable around 9 A.M. (blood pressure peaks then). Births are most likely to occur in the early morning, around 3–4 A.M.; the same is true of deaths, with the curious implication that we all tend to live an exact, whole number of days.

It's a tidy story, except for one loose end: We still haven't explained what's going on when people spontaneously desynchronize, as Siffre did in Midnight Cave. When that happens, the timing of sleep seems to disobey the pacemaker's commands altogether. Can that really be so, or is there another secret hiding in the data, a missing key to the circadian code? This was the problem I dreamed about solving for my Ph.D.

In the fall of 1982, I arrived at Harvard as a new graduate student in applied mathematics. Across the river in Boston, Chuck Czeisler was just starting as a new assistant professor at Harvard Medical School and Brigham and Women's Hospital. I had heard about Chuck while working over the summer with Art Winfree, who had himself done pioneering work on circadian rhythms. Winfree was especially impressed with Czeisler's recent discovery of the law of sleep duration, and showcased it in a review article in *Nature*, one of the world's top scientific journals. I remember how it stunned me. It seemed amazing that despite the vagaries of human psychology and volition, the sleep-wake cycle could obey such a simple and universal pattern. Internal desynchronization might look erratic on the surface, but at a deeper level it was subtly structured. Maybe other laws were waiting to be discovered. The prospect was exhilarating.

I felt like I'd landed in the right place at the right time. Along with the recent addition of Czeisler, the faculty included Richard Kronauer, a mechanical engineer who had developed the leading mathematical model of human circadian rhythms; Martin Moore-Ede, a physiologist with expertise in the circadian rhythms of squirrel monkeys; and Woody Hastings, a cell biologist who had spent 35 years pursuing the molecular mechanisms of the circadian clock. They were all friendly with one another, and cotaught a course at the medical school, attended by a gaggle of their graduate students and postdocs, all hungry for research opportunities.

I met Czeisler on the first day of the course. Tall and in his early thirties, with a Clark Gable mustache, he looked "like a movie star" (according to my mother, when she saw him interviewed on television years later). And more to the point, after his brilliant Ph.D. work, Czeisler seemed destined for academic stardom. Brigham and Women's Hospital gave him an entire floor of the Old Boston Lying-In Hospital for his lab space. When he took me to see it, we were greeted by the sound of jackhammers. Construction workers were busy converting the space into a time-isolation facility, along the lines of what Weitzman had done at Montefiore.

It would be at least a year until Czeisler could run any new studies. But in the meantime, there were plenty of nagging riddles about the existing data. In particular, Winfree kept harping on a fundamental asymmetry: Sleep duration was predictable, but wake duration was not. Even with the benefit of hindsight, no one had found a way to predict how long a desynchronized subject might stay awake. And that meant that half of the sleep-wake cycle was still an enigma.

To begin looking for a law of wake duration, I collected all the data I could find. Czeisler generously shared his old records from Montefiore, plus some data that the French team had sent him. Winfree passed along a few data sets that he had come across. But for the most part, I scoured the scientific literature for published examples of internal desynchronization. Gathering all this information took about a year. Those were the days before digitizers and enlarging photocopiers, so the process was tedious. I'd find a journal article containing a raster plot. Then I'd have a photographer shoot a picture of it and blow it up so I could accurately measure the durations of all the sleep and wake episodes, using a ruler and magnifying glass.

Eventually I compiled a huge database of desynchronized sleep-wake cycles, and began searching it for patterns. I tried plotting wake duration against any prior variable that seemed plausible: the length of the prior sleep episode, or the phase of the body temperature cycle at the moment of waking.

Sadly, those graphs revealed nothing. Later I looked for relationships between wake duration and *two* prior variables. Again, nothing. If a law of wake duration exists, it has remained elusive to this day.

Throughout this fruitless chase, I kept meeting with my new adviser, Richard Kronauer, a silver-haired scientist of great confidence and optimism. He always had time for me, and he loved poring over data—that was fun for both of us. He also had a pet model of how human circadian rhythms worked. That was not as much fun—especially when I would irritate him with the discrepancies between his model and the data I had assembled. He'd raise his voice. My face would redden. Both of us could be stubborn.

One of Kronauer's favorite ideas was that there were two particular times in the circadian cycle when people would not fall asleep. Forbidden zones, he called them. He'd take out his ruler and draw some parallel lines on a raster plot and say, "Look, the subject never falls asleep in this band or that one." I was skeptical—it's easy to find such patterns if you already believe in them. Kronauer was aware of the human proclivity for self-deception, but he insisted that the zones were in specific, consistent places, the same for every subject.

There was no need to bicker. My database could settle the question. If the forbidden zones were real, they should show up as two valleys in the distribution of bedtimes chosen during internal desynchronization. At the opposite extreme, if the subjects were equally likely to fall asleep anywhere in the circadian temperature cycle, the distribution should be flat.

Kronauer was right. When I graphed the relative frequency of sleep onsets as a function of circadian phase, two prominent valleys emerged, each about 2–3 hours wide, centered about 5 hours after and 8 hours before the time of minimum temperature.

While not strictly forbidden, sleep was far less likely to begin in either of these zones. The corresponding clock times could be estimated by applying the conversion formula mentioned previously: The temperature minimum occurs

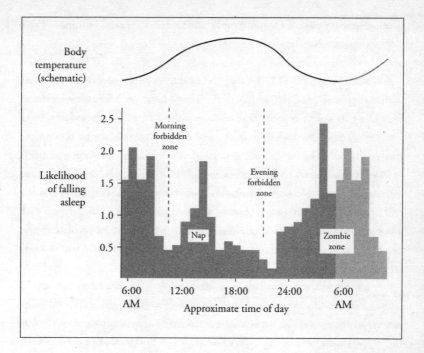

about 1–2 hours before habitual awaking. So for someone who sleeps from 11 P.M. to 7 A.M. each night, the data predicted a "morning forbidden zone" at around 10–11 A.M., and an "evening forbidden zone" at around 9–10 P.M., just an hour or two before bedtime.

The distribution also showed two peaks, representing the sleepiest times in the cycle, in the sense that these were the bedtimes the subjects selected most often (without realizing it, of course, since they were in time isolation). A broad peak centered around the temperature trough coincided with the zombie zone, indicating that this window of minimum alertness was also the time of maximum sleepiness. A second peak occurred about 9–10 hours after minimum temperature, corresponding to siesta time, 2–3 P.M. in the outside world. The intriguing suggestion is that we become sleepy in the mid-afternoon, not because we have eaten a big lunch, or because it is hot outside, but because our circadian pacemaker demands it.

When Kronauer and I saw that the timing of afternoon naps coincided with a sleepiness peak in the desynchronization data, we knew we were on to something. It wasn't obvious that results obtained from time isolation should have anything to say about life in the real world. After all, the conditions are totally different. During entrainment, the rhythms of sleep and body temperature are phase-locked to each other and to the time of day, whereas during internal desynchronization, sleep runs at a different period from temperature, and both run at periods longer than 24 hours. Still, the conversion formula gave a correct prediction of the nap phase, so perhaps we could extrapolate to the rest of the sleepiness rhythm. If so, that meant we should find real-world counterparts of both forbidden zones.

A few weeks later, at a sleep research meeting, I heard a lecture about the hourly distribution of single-vehicle truck accidents, and there it was—the same distribution we had been looking at. (A single-vehicle accident means that the truck jackknifed, overturned, crashed into a bridge abutment, or veered into a ditch of its own accord. It did not collide with another vehicle. The driver probably dozed off at the wheel.) The statistics showed that truck drivers are much more likely to have a single-vehicle accident at 5 A.M. than during regular daytime hours. The second most likely time is between 1 P.M. and 4 P.M., the nap phase. The fewest accidents occur at 10 A.M. and 9 P.M., corresponding to the times predicted for the morning and evening forbidden zones. The explanation seemed clear: Drivers rarely nod off then. Like the zombie zone and the siesta, the forbidden zones must be built into our circadian cycle.

Around the same time, Mary Carskadon, a sleep researcher at Bradley Hospital and Brown University Medical School, was studying the brain waves of subjects on a "constant routine," a protocol designed to unmask the circadian component of the temperature cycle by flattening the subject's behavior and environment as much as possible. Subjects are kept awake in a constant supine posture (propped up in bed) in constant indoor light for 40 hours, and fed hourly portions of a nutrient drink. Although they are supposed to stay awake the whole time, they don't—they occasionally drop off into a "microsleep." For a few seconds the brain falls asleep and the EEG pattern changes suddenly.

Carskadon found that these unintended sleep episodes are more likely to occur at certain times of day. When she graphed the hourly distribution of microsleeps during the last 22 hours of the experiment (by which time the subject was sleep-deprived), she found peaks at the zombie zone and nap phase, and valleys at the two forbidden zones.

Everything was fitting together. The same distribution showed up in microsleeps, in traffic accidents, and in bedtimes chosen during internal desynchronization. Apparently all three reflected the brain's innate circadian rhythm of sleepiness and wakefulness.

Given that the evening forbidden zone is precariously close to habitual bedtime—just an hour or two before it—Kronauer and I wondered if it might be implicated in some forms of insomnia. So far we had no evidence for that. Our time-isolation data showed only that desynchronized subjects seldom *chose* to go to bed about 8 hours before their temperature minimum. The question remained: If people deliberately *tried* to fall asleep then, would they find it difficult?

The answer was already in the literature. In the mid-1970s, several researchers had placed subjects on severely shortened sleep-wake schedules to test their ability to fall asleep at many times of the day and night. For example, Carskadon and William Dement put Stanford college students on a "90-minute day"—a grueling regimen of 30 minutes of bedrest followed by 60 minutes of enforced wakefulness, then back to bed for another 30 minutes, and so on, 16 times a day, around the clock for over five calendar days. Sometimes the subjects made good use of their precious 30 minutes in bed, and fell asleep as soon as their heads hit the pillow. At other times, they could not fall asleep at all, despite being exhausted. Their ability to sleep varied rhythmically each day, and correlated strongly with the phase of their body temperature cycle. The worst time was about 8 hours before the temperature minimum, around 10:30

at night for these college students. Paradoxically, the hardest time to fall asleep was right before their regular bedtime. That surprised Carskadon and Dement, but now it made sense—they had been seeing the powerful effects of the evening forbidden zone.

Kronauer found further evidence that the forbidden zone could cause sleep-onset insomnia. In an experiment conducted at the Cornell Institute of Chronobiology, Jeff Fookson and his colleagues entrained a healthy 21-year-old man to a rigid 23.5-hour day. In other words, while sequestered in a time-isolation facility, he was placed on a strict regimen that involved his going to bed and being awakened a half hour earlier each day. No naps were allowed—only a single, consolidated block of bedrest for 7.75 hours on each cycle. He frequently couldn't take advantage of his time in bed: If he didn't fall asleep immediately, he tossed and turned for about 3 hours. As the experiment dragged on, his sleep deficit grew, yet he continued to suffer from insomnia at bedtime. The subject complained bitterly about the imposed schedule—something was wrong, though he didn't know what—and threatened to quit the experiment. All this from a mere half-hour shortening of his day!

Kronauer's explanation was that the short schedule wrenched the subject's internal phase relations out of whack, anchoring the evening forbidden zone at his scheduled bedtime, making it hard for him to fall asleep. To understand how speeding up the schedule could have this effect, think of the circadian pacemaker as a reluctant dog being dragged around a circular track by a fast-walking owner. (Like the poky dog, this subject's pacemaker tended to dawdle through its circadian cycle every 24.7 hours, whereas the outside world tugged it along impatiently, aiming to finish a lap every 24 hours.) Now if the owner speeds up and walks even more briskly, the dog does too, but it stretches the leash and lags farther behind. For the pacemaker, this means that when the schedule speeds up from 24 hours to 23.5 hours, all events tied to the pacemaker (including the forbidden zones) will similarly lag behind and shift to later times, relative to the schedule. Hence, a forbidden zone that had been safely perched a few hours ahead of bedtime would now be sitting precariously close to it, or even right on top of it. And it would have to stay there, anchored in this awful condition, until something broke the leash and freed it from entrainment. The final part of the experiment supported this interpretation. When the imposed cycle was mercifully shortened to 23 hours, the subject's temperature rhythm broke loose—it could not synchronize to such a short schedule. Consequently his forbidden zone unlocked from his bedtime, his insomnia disappeared, and his mood brightened.

There are people in the real world with the same sort of insomnia as this subject, and for the same reason. He had an intrinsic circadian period of 24.7 hours and had trouble living on a 23.5-hour day; by the same token, people with intrinsic periods near 25.2 hours living in a 24-hour world could well find themselves trying to fall asleep in a forbidden zone. That may be the explanation for the "delayed sleep phase syndrome" estimated to afflict hundreds of thousands of people. Its sufferers are able to sleep well but only at the wrong time of day, like 4 A.M. to noon. This makes it practically impossible for them to hold any job that requires alertness in the morning.

With the evening zone so close to habitual bedtime, even people without sleep disorders may sometimes find themselves trying to fall asleep when it's most difficult. If you've ever gone to bed a few hours early, perhaps because you need to wake up early to catch a plane, you may have noticed how hard it is to fall asleep. The problem is not only that you're excited about the upcoming trip; you're also trying to sleep at the worst time in your circadian cycle. The same thing explains why Sunday night is the worst for insomnia. By staying up late and sleeping in on the weekend, you may have inadvertently allowed your circadian pacemaker and its evening forbidden zone to drift later and possibly intrude on your regular weekday bedtime.

Many people suffer from other forms of deranged synchronization to the 24-hour day, or a lack of synchronization altogether. Shift workers in particular are befuddled by mixed messages. When they're working nights, their circadian pacemakers tell them to sleep during the daytime, but sunlight and traffic noise (and their children) tell them otherwise. In fact, shift work poses major problems for all industrialized societies, problems that will only grow worse. Economics is pushing us to a 24-hour society, with factories and businesses and financial markets operating round the clock. About a quarter of the U.S. workforce already lives on these unnatural schedules. Although the economic benefits are obvious, it's harder to quantify the costs to society and to the workers themselves. These include disrupted family and social lives, gastrointestinal problems, sleep disorders, and the costs of blunders committed while operating in the zombie zone, sometimes with catastrophic consequences.

A candidate for the worst schedule ever is that used by the U.S. Navy on nuclear submarines. The sailors are assigned to 6 hours of duty followed by 12 hours of rest—in other words, they are required to live on an 18-hour day. The pacemaker cannot possibly entrain to such a short cycle, and the sailors live in a perpetual state of desynchronization. The navy's rationale is that an 8-hour shift is too long to maintain vigilance, and there is only room on the sub for 3 shifts of men, hence the 18-hour (3 times 6) schedule. The medical consequences of life on an 18-hour day are unknown, but some indication of the problem is the tremendously high turnover of enlisted men in U.S. submarine crews (about 33 percent to 50 percent per voyage), with only a small number returning for more than two or three of the 90-day missions. Meanwhile, the officers typically live on a 24-hour schedule and tolerate submarine duty much longer, often spending years on active duty.

Sunlight is by far the most important cue for keeping our bodies in sync. Its effect on the pacemaker varies across the circadian cycle, a clever evolutionary design that ensures that the internal clock is always reset in the right direction. Specifically, sunlight in the subjective morning speeds the clock up (as if to tell the body, you missed sunrise today so I'll wake you earlier tomorrow). Sunlight in the middle of the day has little effect on the clock, and sunlight in the evening slows it down. Some correction is needed each day, because the human circadian pacemaker tends to run a bit slow, with a natural period slightly longer than 24 hours. Scientists are still trying to work out exactly how light entrains the pacemaker, but in outline, we know that light strikes the eyes and produces a chemical change in photoreceptors in the retina, which then forward an electrical signal along neural pathways to the suprachiasmatic nuclei in the hypothalamus, the site of the pacemaker. Surprisingly, the photoreceptors have not yet been identified. They are not the rods and cones we use for vision; blind mice with a genetic disorder that destroys their rods and cones can still entrain to a light-dark cycle.

Further evidence of the synchronizing effect of light can be inferred from the fact that 80 percent of blind people have chronic sleep disorders. Unable to reset their clocks by the necessary amount each day, they have trouble sleeping or staying alert at the socially appropriate times. Their complaints are periodic; for two or three weeks of every month, when they're out of step with the world, their daytimes are riddled with uncontrollable naps, their nighttimes plagued by fractured sleep. But gradually their biological clocks drift so late that they come back into harmony with the rest of society. Then they feel fine for a week or two before the next wave crashes in.

Remarkably, the other 20 percent of blind people do manage to synchronize to the light-dark cycle. The likely explanation is that the circadian photoreceptors in their retinas are intact, even if their rods and cones are not. This allows light to work its resetting action on the clock, by striking the eyes and then traveling down the neural pathways to the pacemaker. In other words, although these people lack sight, they can still perceive light in a nonvisual, circadian sense. The evidence for this surprising idea comes from recent studies involving melatonin, a brain hormone produced by the pineal gland. In sighted people, the secretion of melatonin ebbs and flows on a daily cycle, with peak output in the dark of night while we're asleep. This circadian rhythm is driven by the master clock, just like body temperature, alertness, and so many other physiological functions. In that sense, melatonin levels provide another proxy for the pacemaker. Furthermore, the secretion of melatonin is responsive to light—it plummets whenever bright light enters the eyes. (Here, "bright" means light of typical daytime intensity, much brighter than typical indoor light but otherwise nothing extraordinary.) In 1995, Czeisler and his colleagues tested the melatonin suppression response of totally blind subjects by exposing them to bright light at a time when the melatonin levels in their blood were high. Most subjects showed no suppression at all, as one would have expected: The light was not getting through to their clocks. But among that special subpopulation of blind people who somehow manage to sync to the 24-hour day, the light turned off the melatonin secretion, just as it does in normally sighted people. The implication is that there are two pathways from the eyes to the brain: one for conscious vision and the other for circadian entrainment. This hypothesis is consistent with the known anatomy of the mammalian brain; the neural hotline to the pacemaker is separate from the brain's visual pathways.

Just as blind people are teaching us about the nature of the circadian photoreceptor, a population affected by a different syndrome is illuminating the inner workings of the clock itself. Scientists have recently found the first gene linked to a human circadian rhythm by studying patients with a rare disorder, discovered in 1999, called "familial advanced sleep phase syndrome." Family members afflicted by it are extreme morning-types, falling asleep at around 7:30 P.M. and waking spontaneously at 4:30 A.M. Lab studies showed that the circadian clocks in these people run fast, with a period about an hour shorter than normal, suggesting a genetic mutation in clock function. A research team at the University of Utah led by Louis J. Ptacek traced the mutation to a single gene, hPer2, whose protein product is believed to play an essential role in the molecular feedback loops that generate circadian oscillations in single cells.

Some other families with the syndrome do not possess mutations of that gene, which means that some other mutant genes probably exist. Once enough mutants are available, we can expect scientists to make rapid progress in dissecting the molecular and genetic basis of human circadian rhythms. And that will inevitably lead to more effective treatments for jet lag, shift work, and sleep and psychiatric disorders associated with derangements of daily sync.

II

DISCOVERING SYNC

· Four ·

THE SYMPATHETIC UNIVERSE

THE SCIENCE OF SYNC HAS COME A long way since Androsthenes, the scribe for Alexander the Great, gave the first written description of a biological rhythm. Around the fourth century B.C., while on the march to India, he observed that the leaves of tamarind trees always opened during the day and closed at night. It would take another two millennia before mankind would stumble across an eerier kind of sync, the synchronization between things that aren't even alive.

Some of the pivotal discoveries in the history of science were made by serendipity. Think of Alexander Fleming, who, as we all know, discovered penicillin when an airborne mold contaminated his experiment and killed the bacteria he was studying. Or take Arno Penzias and Robert Wilson, scraping the pigeon droppings off their giant radio antenna at Bell Laboratories, trying to eliminate the annoying background hiss that seemed to be coming from outer space in every direction, until they realized that they were hearing the birth cry of the universe, the 14-billion-year-old echo of the big bang.

Although the role of serendipity is familiar, what's not so well appreciated is how different serendipity is from luck. Serendipity is not just an apparent aptitude for making fortunate discoveries accidentally, as my dictionary defines it. Serendipitous discoveries are always made by people in a particular frame of mind, people who are focused and alert because they're searching for something. They just happen to find something else.

So it was with the discovery of inanimate sync. In February of 1665 the Dutch physicist Christiaan Huygens was confined to his bedroom for several days, ailing with what he delicately described, in a letter to his friend Sir Robert Moray, as a "slight indisposition." He'd fallen behind on his correspondence— he owed Moray three letters—and now he was writing with the news of a strange phenomenon he'd observed while cooped up in his room, a "marvelous thing which will surprise you."

In the room with him were two pendulum clocks—the two most accurate timekeepers ever built. Huygens had invented the pendulum clock a decade earlier, and now, with its help, he hoped to solve the greatest technological challenge of his day: the problem of determining longitude at sea. As beautifully recounted by Dava Sobel in her best-selling book *Longitude,* a solution to the longitude problem took on the gravest importance in the Age of Exploration, as more ships sailed across the oceans to trade with other nations, to wage war, and to conquer new territories. Unlike latitude, which measures a ship's angular distance from the equator and which is easily gauged from the length of the day or the height of the sun above the horizon, longitude—the ship's angular position east or west on the globe—is defined arbitrarily, with no intrinsic counterpart in the environment. Sailors could not use the stars or sun or any other physical cues to determine their longitude, even aided by the best charts and compasses. Without any way to establish their whereabouts at sea, even the finest captains lost their way and drifted hundreds of miles off course or ran aground on rocky shores. Those keeping to familiar routes were easy prey for pirates. The governments of Portugal, England, Spain, and Holland offered vast rewards for a workable solution. Although the puzzle was tackled by some of the leading scientists of the era—Galileo, Giovanni Domenico Cassini, Isaac Newton, Edmond Halley—it remained unsolved for over four centuries.

Now Huygens was pursuing a solution along the lines originally suggested by the Flemish astronomer Gemma Frisius, who realized in 1530 that longitude could be determined, at least in principle, by accurate timekeeping. Suppose a ship had an onboard clock that was set correctly upon departure from the home port, and that always ran true thereafter. By carrying "home time" out to sea in this way, a navigator could determine longitude by consulting the clock at the exact moment of local noon, when the sun is highest in the sky. Since the Earth takes 24 hours to complete 360 degrees of rotation, each hour of discrepancy between local time and home time corresponds to 15 degrees of longitude. In terms of distance, 15 degrees translates to a thousand miles at the equator, so for this strategy to be practical, the clock had to be accurate to within a few seconds a day. The challenge was to devise a mechanical clock that never wavered, despite the heaving of the ship on violent seas, and despite the assaults of ever-changing humidity, pressure, and temperature, which can rust a clock's gears, expand its springs, or thicken its lubricating oil, causing it to speed up, slow down, or stop.

All clocks of the 1500s and early 1600s were woefully inadequate. The best of them would lose or gain fifteen minutes a day, even under ideal conditions. Huygens's new pendulum clocks, however, were a hundred times more accurate. A solution to the longitude problem finally appeared to be within reach. On a sea trial in 1664, conducted in partnership with the Royal Society of London (and a cooperative captain), two of Huygens's specially designed maritime clocks sailed to the Cape Verde Islands, off the west coast of Africa, and successfully tracked the longitude thoughout. The two-clock design provided useful redundancy; in case one of the clocks stopped or needed to be cleaned, the other could still keep accurate time. Unfortunately, later tests revealed the clocks to be temperamental. They performed well in favorable weather, but stormy seas bothered the swinging of their pendulums.

Meanwhile, Huygens remained in The Hague and corresponded with the Royal Society through Sir Robert Moray, both to inquire about the results of the ongoing sea trials, and to report on his latest attempts to perfect the design of his clocks. It was around that time, on a quiet day in late February 1665, that serendipity struck. In a letter to his father, Huygens wrote:

> Being obliged to stay in my room for several days and also occupied in making observations on my two newly made clocks, I have noticed an admirable effect which no one could have ever thought of. It is that these two clocks hanging next to one another separated by one or two feet keep an agreement so exact that the pendulums always oscillate together without variation. After admiring this for a while, I finally figured out that it occurs through a kind of sympathy: mixing up the swings of the pendulums, I have found that within a half hour they always return to consonance and remain so constantly afterwards for as long as I let them go. I then separated them, hanging one at the end of the room and the other fifteen feet away, and noticed that in a day there was five seconds difference between them. Consequently their earlier agreement must in my opinion have been caused by an imperceptible agitation of the air produced by the motion of the pendulums. The clocks are always shut in their boxes, each weighing a total of less than 100 pounds. When in consonance, the pendulums do not oscillate parallel to one another, but instead they approach and separate in opposite directions.

In a letter to his friend R. F. de Sluse on February 24, 1665, Huygens described the sympathy effect as "miraculous." On February 27, he dashed off the letter to Moray, asking him to convey the observations to the Royal Society.

Over the next week Huygens conducted a series of experiments to explore what might be causing the sympathy. He hung both clocks from hooks embedded in the same wooden beam, and found that when he turned them at 90 degrees to one another, or separated them by more than 6 feet, their sympathy disappeared. Yet when he placed a large board between them to block any passage of air, the sympathy persisted. So his first guess was wrong; the clocks were not communicating through the air after all.

He then suspected that the clocks might be interacting through tiny vibrations of their common support. To investigate this possibility, he tried hanging each clock from a separate plank, with both planks lying on top of two rickety chairs positioned back to back. Again the clocks sympathized. Their pendulums swung apart and together, apart and together, like a pair of hands clapping. When one clock sounded *tick*, the other sounded *tock*. Then he disrupted their sympathy to see what would happen. The result must have spooked him—the chairs began to shake. In sympathy they had been motionless, but now they were trembling, clattering on the floor. They continued to shake for another half hour until the sympathy restored itself, at which point the chairs fell silent.

Huygens had his answer. Even though each clock was housed in a heavy box weighted with 80 or 90 pounds of lead, the swinging of its pendulum imparted a slight movement to the box, which jiggled the planks, which jiggled the chairs. But when the clocks were in sympathy—when their pendulums swung precisely opposite to each other—the equal and opposite forces they exerted on the planks canceled each other out, which allowed the chairs to keep still. Conversely, when he disrupted the sympathy, the opposing forces no longer balanced at all times. A portion of them added up and dragged the planks back and forth from side to side, shaking the chairs. As Huygens put it, "Once the consonance is achieved the chairs will not move any more, only preventing the clocks from leaving [the state of sympathy], since as soon as they try to do that, the small movement of the chairs restores them to the previous position." In modern terms, Huygens had just invented the concept of stabilization by negative feedback.

The Royal Society was disappointed by this explanation, not because they thought it was wrong, but because they feared it was right. The minutes of the meeting of March 8, 1665, record that "occasion was taken here by some of the members to doubt the exactness of the motion of these watches at sea, since so slight and almost insensible motion was able to cause an alteration in their going." In other words, Huygens's own reasoning suggested that his pendulum clocks were exquisitely sensitive—and therefore too sensitive to solve the longitude problem.

The sympathy of clocks, which seemed so miraculous just a few weeks earlier, now struck Huygens as a nuisance. He never explored it again, nor did he ever manage to solve the longitude problem. Its solution had to wait another hundred years. In the mid-1700s, John Harrison, an Englishman with no formal education, developed a series of maritime clocks made of various metal parts that resisted rust and ingeniously compensated for one another's expansion and contraction at different temperatures. His fourth chronometer, the masterpiece he called the H-4, contained jeweled parts made of diamond and ruby, which enabled it to run almost friction-free. It weighed only three pounds, with a diameter of five inches—no larger than a big pocket watch. When tested at sea in the 1760s, it tracked longitude to an accuracy of 10 miles, sufficient to win the British Parliament's prize of 20,000 pounds, equivalent to a few million dollars today.

Ironically, as the longitude problem recedes into history, the sympathy of clocks grows more central to science with each passing year. As great a genius as Huygens was ("Summus Hugenius," Newton called him), even he could not grasp the full significance of what the universe had disclosed that day in his room. But with more than 300 years of hindsight, we can see it. Huygens had discovered one of the most pervasive drives in all of nature. Huygens had discovered inanimate sync.

We take it for granted that we can sing and dance together, march in step, clap in unison. Sync is second nature to us. But because it comes so easily, we have poor insight about what it actually demands. It seems to involve at least a low level of intelligence, the ability to time our behavior and anticipate that of others. Which is why the reports of concerted flashing among thousands of fireflies aroused such skepticism for so many years, and why we are impressed by the chorusing of crickets or the seductive tactics of male fiddler crabs, who court a female by waving their gargantuan claws at her in unison.

Still, these feats of living sync can always be chalked up to the marvels of evolution, the magic of millions of years of natural selection. And it's in that light that we can see most clearly what was shocking about Huygens's serendipitous discovery.

His pendulum clocks were not alive.

Mindless, lifeless things can sync spontaneously.

The sympathy of clocks taught us that the capacity for sync does not depend on intelligence, or life, or natural selection. It springs from the deepest source of all: the laws of mathematics and physics.

That insight has led to a great flourishing of sync in technology. For example, without sync, we wouldn't have laser eye surgery, or CD players, or supermarket checkout scanners, or any of the other everyday wonders where lasers are used. The intense, coherent, needle-thin beam of a laser is a result of trillions of atoms emitting light waves in sync. The atoms themselves are no different from those in an ordinary lightbulb—the trick is in the way they cooperate. Instead of cacophonous light of different colors and phases, laser light is one color and one phase, like a chorus singing the same note. It can be made very intense (though it doesn't have to be); it travels in a narrow beam; and it can be focused on a tiny spot. In contrast, ordinary light can be made intense only at the cost of a prohibitive amount of energy; it spreads out and weakens rapidly as it travels; and it is difficult to focus. All these advantages of laser light allow it to be controlled and manipulated easily. Surgical lasers, for example, produce a point of concentrated energy that's smaller than any scalpel and can therefore reach diseased tissues that would otherwise be inaccessible. Furthermore, there's much less bleeding with laser surgery, because clotting occurs instantly; the laser cauterizes the incision as it cuts.

For years after the invention of the laser, no one knew what to do with it. Some teasingly described it as a solution looking for a problem. And yet this child of basic research, born of pure curiosity about light waves in sync, has become one of the most versatile devices of our time, with applications that no one could have foreseen. At a party celebrating its fortieth birthday, Arthur Schawlow, cowinner of the 1981 Nobel Prize in Physics (in part, for developing the laser along with Charles Townes) recalled:

> We thought it might have some communications and scientific uses, but we had no application in mind. If we had, it might have hampered us and not worked out as well. . . . It's nice that there are medical uses. Some of you have probably heard me say before that although there is a lot of talk in the

newspapers about death rays, there still aren't any real death rays as far as I know. But one of the first applications of lasers was for surgery of the retina in the eye to prevent blindness from retinal detachment. Neither Charlie nor I had ever heard of surgery for detached retinas to try to prevent blindness, and if we had, we probably wouldn't have been fooling around with stimulated emission from atoms.

That phrase—"stimulated emission from atoms"—is the secret of how a laser works. But I'm a little embarrassed to admit, I've had lasers explained to me about ten times, and the explanation never seems to stick. All the talk of excited atoms and population inversions goes in one ear, lingers for a few seconds of hazy understanding, and then seeps out. I keep hoping to find a simple analogy that will make sense to me, something I could picture and remember more easily, and now I think I've finally come up with one—but it's pretty crazy. If you already understand lasers, or if you don't really care about how they work, feel free to skip to the next section.

Imagine you wake up one morning and find yourself on an alien planet, entirely deserted except for a watermelon with a step stool beside it. Naturally you wonder what the stool is for, so you take a guess and place the watermelon on top of it. The melon becomes strangely agitated, fidgeting on its perch. Almost immediately it rolls off and crashes to the ground. At the same instant, it spits out a seed like a bullet, flying off in a random direction.

What I've described so far is an analogy for the way that ordinary light is produced. Say you turn on your toaster and the coil glows bright red. What's going on here is that electricity is flowing through the coil and heating it up. The heat raises the atoms in the coil to a higher energy level, which is what lifting the watermelon onto the stool is supposed to represent. After a very short time, each hot atom spontaneously falls back to its lowest energy level—its "ground state"—and gives up its excess energy by emitting a photon (a particle of light) in a process called spontaneous emission; this is like the fidgety watermelon rolling off the stool and shooting out a seed. So a hot coil glows red because its excited atoms are spontaneously emitting a lot of red photons.

As you continue to explore the planet of the watermelons, you soon come to the edge of a vast field, with millions of watermelons lying on the ground, each with its own stool next to it. This raises an intriguing new possibility. What would happen, you wonder, if a seed-bullet happened to strike another watermelon? To start the action, you lift one melon onto its stool. It soon falls off and fires a seed in a random direction, and by good luck, its flight path carries it smack into another melon lying on the ground. As soon as the target melon absorbs the impact, it jumps up onto its own stool, where it too starts to quiver, and before long, it drops down and fires a seed of its own, again in some random direction. It's an amazing spectacle, one seed triggering another, melons jumping up onto stools and then falling back down. By lifting the initial watermelon, you have inadvertently created a chain reaction, although of a very feeble, nonexplosive sort: Its size stays fixed at just one seed in the air at all times. Actually, if a seed ever fails to hit a watermelon and flies out of the field, the process stops altogether.

This cascade process is interesting, but it's not the analog of a laser. It doesn't amplify light; it never increases the number of photons in the air. The missing piece has to do with the only aspect of watermelon physics we haven't considered yet: What happens if a seed hits a watermelon while it's tottering on its stool, instead of one lying calmly on the ground? To find out, you lift many watermelons onto their stools at the same time, running quickly from melon to melon before any of them has a chance to roll off. Then you stand back and watch. Eventually a melon drops down spontaneously and fires a seed, scoring a direct hit on another melon wobbling on its stool. (The odds of this are good, because you've lifted so many melons onto their stools ahead of time.) What happens next is astonishing. Instead of being absorbed, the incoming seed passes straight through the melon without changing its flight path; what's even weirder, it is now accompanied by another seed exactly like itself, moving in tandem with it. In effect, the incoming seed has been cloned. Before, there was one seed flying in that direction; now, there are two.

This is the key process behind a laser. It's called stimulated emission, and you can see that it offers a way to increase the number of photons flying along a certain line. Every time a photon hits an excited atom, it duplicates itself, amplifying the amount of light traveling in that direction, which is precisely what the acronym *laser* stands for: Light Amplification by Stimulated Emission of Radiation. The emission is said to be stimulated (as opposed to spontaneous) because the incoming photon provoked the excited atom into spitting out the new photon.

What matters most, however, is that the emitted photon is indistinguishable from the one that spawned it. If you think of these photons not as particles but as tiny waves of light, they'd be perfectly synchronized. All their peaks and valleys would be aligned, meaning that they're carrying light of the same color, in the same direction and with the same phase.

There's no commonsense way to understand how stimulated emission could be possible, or why the new photon should be a carbon copy of the old one. The phenomenon is a consequence of the odd logic of quantum mechanics, the physics of the atomic and subatomic world, where our intuition from everyday life breaks down. Einstein discovered the theoretical necessity of stimulated emission in 1917, but it took another 43 years before anyone figured out how to use it to create the first working laser.

Actually, stimulated emission is not enough; lasers rely crucially on two other ingredients. First, we have to find a way to keep most of the watermelons on their stools for most of the time, since they are the only targets that can give rise to stimulated emission. Watermelons on the ground are useless. And that means we have to invest a lot of energy, since the watermelons drop back to the ground every time a stimulated emission occurs. The process of continually lifting them back up is known as "pumping" the laser to create a "population inversion." Depending on what type of laser you're using, you excite the atoms simultaneously by heating them, or blasting them with a flash lamp, or sending electrical discharges through them. That injection of energy inverts the population, in the sense that it hoists a large fraction of the atoms up to a higher energy level than their preferred spot in the ground state.

The second thing that's needed is a way to intensify the light, and to create a narrow beam moving in a single direction. Both are achieved by placing the atoms in an echo chamber for light, or what a physicist would call a resonant cavity. An organ pipe is a resonant cavity for sound. So is the body of a guitar; it amplifies the faint vibrations of a plucked string into the full sound of the instrument. A laser's cavity does the same thing, except with waves of light. Take a long, thin glass tube and fill it with a gas of the right kinds of atoms or molecules, or take a solid rod of ruby; there are many ways to make a laser. Then put mirrors at both ends. Flip the switch to begin pumping the laser (lift those watermelons). Spontaneous emission starts the chain reaction. Remember, those first photons are ejected in random directions. Then, when they trigger the subsequent process of stimulated emission, those initial photons clone themselves, but since they are still moving in whatever random directions they started with, many of them bang into the walls of the tube and get absorbed; they do not contribute to the laser light. In other words, all those directions have now been neatly filtered out. Only the photons bouncing back and forth between the mirrors survive. And not only do they survive; they proliferate. With every rebound through the tube, they give birth to more and more perfect copies of themselves, reinforcing their light and creating a magnificent beam of perfectly synchronized photons. To let some of that light out, one of the mirrors is designed to be slightly less than 100 percent reflective. The tiny fraction of synchronized light that escapes is what we see as a laser beam.

The central mystery here—why the newly created photons are always in sync with the ones that made them—will come up again in the next chapter, when we take a deeper look at synchrony in the quantum realm.

Another kind of synchrony lies at the heart of the American power grid, the electrical behemoth that supplies alternating current to the outlets in your home and office. Thousands of generators in power plants all across the country are linked together to make two gigantic, synchronous machines, the regional power grids that serve all the states east or west of the Rocky Mountains. (Texas, being Texas, has its own.) Each grid functions like a single enormous generator, with all its component generators rotating in unison.

I'd heard of the power grid for years without ever really considering what it meant. And maybe, like me, you have never given much thought to where your electricity comes from, or if you have, you supposed it was generated nearby at the local power plant, and that the same was true for everyone else. The truth, however, is that during a heat wave in the Midwest, an air conditioner in Wisconsin might actually be running on power generated a moment earlier at a plant in South Carolina. Without sync, that seamless transfer of power wouldn't be possible.

In outline, the system works like this. Each power plant harnesses some form of natural energy to drive a turbine that spins a generator that produces electricity. For example, the plant might burn coal, oil, or natural gas, or use nuclear energy, to create enough heat to boil water into steam, and then use the steam to rotate the turbine. Or it might use the energy of flowing water (like at Niagara Falls) to turn a hydroelectric waterwheel. Once the electricity has been generated, it is transformed to the much higher voltages (up to 765 thousand volts) used to send it on the nationwide transmission grid. This allows plants to ship electricity cross-country to compensate for shortages elsewhere or to exploit price differentials. At the end of the line, the electricity is stepped back down to the 120-volt power we use in our homes and offices.

The origins of the grid can be traced back to 1882, with the opening of Thomas Edison's Pearl Street Station in Manhattan, which served electrical power to 59 customers. The new technology was an instant sensation, and by the late 1880s, several other cities were electrified. Edison's young company, General Electric, provided the kind of electricity known as direct current (the familiar kind that a battery supplies), in which current flows steadily from high voltage to low voltage, analogous to water flowing downhill.

The trouble with direct current, unfortunately, was that it couldn't be transmitted more than a few city blocks. On longer journeys, too much of the power was lost to heat, the inevitable consequence of the resistance in the wires. The only remedy was to transmit electricity at very high voltage and very low current (because the wasted energy grows in proportion to the square of the current, so it's best to keep the current as low as possible). But that was not an option because Edison's customers needed *low* voltage, not high, to run their little lamps and primitive gadgets. What was desperately needed was a device called a DC transformer, something that could convert direct current from high voltage to low voltage. No one at the time, not even Edison, could figure out how to make one.

Meanwhile, in the 1890s, the Westinghouse Company was experimenting with a new kind of electricity called alternating current, pioneered by Nikola Tesla, in which the current alternately reversed its flow direction in sync with the rotation of the spinning generator that produced it. After acrimonious debates about the merits of the two approaches, alternating current won out because it was much easier to transform from high voltage to low voltage and back again. Also, the generators were inherently simpler, because rotating magnets automatically create alternating current, whereas an extra step is needed to change it to direct current.

The main question about alternating current was what frequency to use. In other words, how many times per second should the current swing back and forth? In 1900, when the decision was up for grabs, many of the local electric utilities operated independently and tried different choices. Some stuck stubbornly to direct current, while others produced alternating current at 25, 50, 60, 125, or 133 cycles per second. For instance, the power plants at Niagara Falls and other hydroelectric stations favored 25 cycles per second, because the turbines in the generators could be designed to run more efficiently at those speeds. That frequency had a curious drawback, not on engineering grounds, but on psychological ones: It caused incandescent bulbs to flicker at a rate that most people found noticeable and disturbing. (Today, the standard frequency for alternating current in North America is 60 cycles per second, while 50 cycles per second is common throughout the rest of the world.)

Gradually, as the demand for electrical power grew, the local utilities expanded and encroached on one another's territory. It was around this time that the interconnected power grid was born. Consolidation offered several advantages. A networked system was more reliable, because one plant could pick up the slack if another had an equipment failure or a shortfall of generation capacity. There was also an economic benefit: Utilities in different regions could buy and sell one another's power, exploiting disparities in the cost of service. Sometimes it was cheaper to buy power off the grid than to generate it yourself.

A technical difficulty with interconnecting was that all the generators had to be synchronized to spin at exactly the same rate, even though they might be separated by hundreds of miles. Synchrony was crucial. Without it, power would slosh back and forth through the grid, causing tremendous current surges in the transmission lines. In the worst case, a generator might draw so much power that it could explode or be severely damaged. (Today, special protective equipment disconnects any generator that falls out of step.) Part of the solution came from the laws of physics. Electrical engineers found that generators connected in parallel had inherent tendencies to synchronize their rates of rotation. In other words, a parallel grid tends to be self-synchronizing: a beautiful instance of spontaneous sync, in the spirit of Huygens's sympathy of clocks.

The effect is easiest to understand for the case of two generators connected in parallel. If they ever happen to rotate at different speeds, the slower generator automatically draws power from the faster one, so the slower one speeds up and the faster one slows down, which corrects the discrepancy. In more physical terms, any disturbance that causes one generator to pull away from the other is opposed by corrective electrical currents that immediately begin circulating, which set up torques that cause the speeds of the generators to become more nearly equal. Thus the pair of generators tends to synchronize spontaneously.

The downside of interconnectivity is that failures can propagate. These domino effects can be complex, unpredictable, and dramatic. During rush hour on the night of November 9, 1965, the high-voltage power lines from Niagara Falls to New York City were running at maximum capacity when a torrent of electrical energy went on a rampage. Shortly before 5:15 P.M., a protective device malfunctioned and choked off 300,000 kilowatts that were supposed to be headed for New York City and instantly forked them elsewhere on the grid, triggering a chain reaction in which one circuit breaker tripped after another, splitting the entire Northeast power system into disconnected electrical islands. Toronto went black at 5:15, Rochester at 5:18, Boston at 5:21. Ultimately 30 million people in New Hampshire, Vermont, Massachusetts, Connecticut, Rhode Island, New York, metropolitan New York City, and parts of Pennsylvania lost their power, some for up to 13 hours.

It's understandable that cascading failures like this should happen occasionally. The power grid is an enormously complex dynamical system. Its job is formidable: to provide electricity on demand, instantaneously, and at the correct voltage levels and frequencies. Unlike other products, electricity can't be stored. It has to be produced on the spot; power generation is the original "just in time" industry. Complicating the task immensely is that the demands on the system depend on uncontrollable factors, like heat waves or quirks of human psychology. When the verdict in the O.J. Simpson trial was read, the entire power grid sped up from a sudden drop in consumption, presumably because millions of people turned off their television sets simultaneously as soon as they heard the verdict. Now, with the deregulation of the power industry, and the potentially destabilizing impact of free-market economics on the functioning of the grid, engineers and scientists will face even greater challenges in ensuring that the largest machine ever built continues to function as reliably as it has for decades.

In other technological settings, sync is used to keep things organized. Precise agreement on the time of day at two or more remote locations is crucial for electronic bank transfers, for synchronizing television feeds, and for transmitting everything from E-mail to the songs on the radio. (When you tune in to a radio station, you have to set the dial to the right frequency, which enables your radio to sync with the broadcast. Without that, you wouldn't be able to home in on the radio wave carrying the music, and you wouldn't hear anything but static.) The same principle is used in cell phones and satellite communications, and all other forms of wireless communications.

All the electrical components on a computer chip are clocked to operate in sync. A microelectronic crystal beats billions of times each second, switching the digital circuitry on and off in concert, which helps the millions of circuits on the chip communicate with one another efficiently. This centralized design, with all components slaved to a tyrannical master clock, has some notable disadvantages: 15 percent of the circuitry is wasted on distributing the clock signal, and the clock itself consumes 20 percent of the power. But engineers still favor this design because of its conceptual simplicity, and because the alternative—a democracy of many local clocks, as in firefly swarms and circadian pacemaker cells—is still not well enough understood to be easily imitated in practice.

The most high-tech applications of sync are direct descendants of Huygens's pendulum clocks and the longitude problem. Today the world's best timekeepers are devices known as atomic clocks. Like all earlier clocks, they rely on counting the oscillations of a periodic event. But instead of the rising of the sun, or the dripping of a faucet, or the back and forth swings of a pendulum, atomic clocks count the transitions of a cesium atom as it flits back and forth between two of its energy levels. The universal time standard, the NIST-F1 maintained by the National Institute of Standards and Technology in Boulder, Colorado, is a cesium superclock that errs by less than a second in 20 million years. A new optical clock is under development that will be a thousand times better still. It wouldn't have lost a second since the universe began.

The obsession with keeping accurate time is more than a sign of scientists' fastidiousness. Just as accurate clocks were the key to solving the longitude problem, atomic clocks have made it possible to pinpoint a location anywhere on Earth to within a few meters. The technology is known as the global positioning system (GPS). Developed by the American military to allow ballistic missiles to be launched more accurately from submarines, the global positioning system first came to the public's attention in 1991, when it guided cruise missiles through windows in Baghdad, and enabled coalition troops to find their way in the Iraqi desert at night. Peacetime applications range from helping lost drivers in rental cars, to precision farming and enhanced 911 systems that automatically calculate the fastest routes for ambulances and fire trucks. Refined versions of GPS are being tested for blind landing of airplanes in heavy fog, where the aircraft will need to be positioned to within 10 centimeters both horizontally and vertically. But GPS is more than a navigation system: It allows time synchronization to better than a millionth of a second, which is useful for coordinating bank transfers and other financial transactions.

The global positioning system consists of 24 satellites orbiting about 11,000 miles above the Earth, arrayed so that any spot on the planet is visible to at least six of the satellites at any time. Each satellite carries four atomic clocks on board, synchronized within a billionth of a second of one another by the master superclock in Boulder. Any GPS receiver, like those found in expensive cars or on handheld devices, receives signals from four of these satellites (at least), and uses those four numbers to calculate its three-dimensional location and the current time. The calculation works on a form of triangulation: The satellites emit radio signals continuously, each timestamped to the nanosecond (that's where the onboard atomic clocks come in); the receiver then compares the time of reception to the time of transmission, and multiplies the difference by the speed of light to calculate the distance to the satellite. By doing the same calculation simultaneously with at least four satellites (all of whose positions are known very accurately), the receiver can pinpoint its location to a few meters in less than a tenth of a second.

The power of inanimate sync reaches out into the vastness of space, far beyond the man-made satellites of the global positioning system. We tend to be unaware of sync at a cosmic scale because of the unfathomable distances and times involved. But when astronomers recently discovered two little planets orbiting the star Gliese 876, about 15 light-years away from Earth, one of the first things they noticed was that the planets are locked in orbital resonance, a graceful dance in which one planet goes around its star twice in the same time that the other goes around once. Something even more remarkable happens with our own moon: It turns on its axis at precisely the same rate as it orbits the Earth, which is why we always see the same side of the moon—the one with the man's face on it, not the dark side on the back of his head.

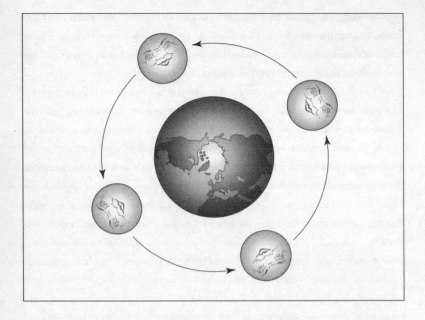

The synchronization between the moon's orbit and its spin can be explained intuitively. To keep things simple, suppose that the moon follows a circular orbit around the Earth. The size of the circle is determined by a balance of two forces: the force of gravity from the Earth pulling on the moon, and the centrifugal force from the moon's motion, which tends to make it fly away from the Earth. (Centrifugal force is the force that pushes you against the door in your car when you race around a tight turn.) The two forces, gravitational and centrifugal, balance each other perfectly at the center of the moon. But keep in mind that the moon is a huge ball, not a point. At points other than the center, the forces are not quite in balance. On the near side of the moon, gravity is stronger; on the far side, centrifugal is stronger. This imbalance creates two small bulges in the moon, one on the near side and one on the far side. The same thing happens on Earth, due to the moon's gravity: This is what causes the tides in the ocean. On the moon, where there is no water, the "tidal effect" is less visible, but important nonetheless, because it deforms the moon from spherical to slightly cigar-shaped. Because of the gravitational pull of the Earth, the cigar always wants to point directly toward the center of the Earth. For that alignment to persist even as the moon orbits the Earth, the moon has to spin at the same time, and in a very precise way—it needs to turn on its axis exactly once for each revolution it makes about the Earth. And that is the condition the moon finds itself in today: a condition known as 1:1 spin-orbit resonance, or tidal locking.

If the moon were ever to depart from this resonant condition, the tidal force would twist it back into alignment. To see why, suppose the cigar were not pointing toward the center of the Earth.

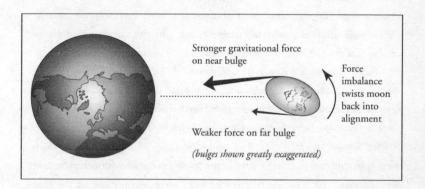

Stronger gravitational force on near bulge

Force imbalance twists moon back into alignment

Weaker force on far bulge

(bulges shown greatly exaggerated)

The situation would then be somewhat like a compass needle that's not pointing north—the force field (magnetic for the compass needle, gravitational for the moon) exerts a corrective torque that tends to restore the cigar to its equilibrium position. Specifically, the Earth's gravity twists the near bulge of the moon in one direction, and the far bulge in the opposite direction, but the near bulge is twisted more strongly because it's closer. The effect is to realign the cigar, thus enforcing the 1:1 spin-orbit resonance.

Instead of a compass needle, an even closer analogy would be to a popular toy from my childhood, a bottom-heavy dummy with a rounded base; if you try to tip "Joe Palooka" over, he rights himself automatically. The moon is bottom heavy in the same way, in the sense that its near bulge is weighted more strongly by the Earth's gravity, thus providing the corrective torque necessary to pull the moon back into synchrony.

Another form of astronomical sync may have been involved in the extinction of the dinosaurs, an event that changed the course of life on Earth forever, allowing small mammals to prosper and evolve into us. According to the reigning theory proposed by the father-son team of Luis and Walter Alvarez and their colleagues, the dinosaurs and many other forms of life were suddenly annihilated when a giant object of some sort—perhaps an asteroid, perhaps a comet—smashed into the Earth about 65 million years ago. With the destructive power of 100 million hydrogen bombs, it created worldwide devastation in the form of wildfires, sweltering temperatures, poisonous acid rain, and impenetrable clouds of dust and smoke that blocked all sunlight for months.

To see how such a cataclysm could possibly be connected to sync, we first need to understand why rocks occasionally fall from the sky and strike our planet. These meteors are thought to be the leftovers of an aborted attempt to form a planet in the early days of our solar system. Back then, particles of dust swirled around the infant sun and gradually coalesced into boulders, which in turn agglomerated into larger and larger pieces, eventually forming the planets we see today.

One of the most striking features of the resulting solar system is the void that separates the inner planets (Mercury, Venus, Earth, and Mars) from the next one farther out, the giant planet Jupiter. Most of us have no sense of how far apart the various planets are. The distances seem incomprehensible. But we're starting to get a feel for them here in Ithaca, thanks to a scale model of the solar system called the Sagan Walk, erected in honor of the late Carl Sagan, who spent much of his career at Cornell. Walking around our town commons, starting at the sun in the middle of the plaza, you immediately encounter the four inner planets, each about the size of a small pea, mounted inside their own Plexiglas displays. It takes just a few steps to walk from one to the next, and you start to realize that it's only a short stroll from Mercury to Mars: All the inner planets are right there on the same plaza. But to reach the next one, you have to leave the commons and walk down the street for a few minutes over to Moosewood Restaurant, where Jupiter awaits. Why the big void between the inner planets and Jupiter?

Actually, it's not a void. Between Mars and Jupiter lies a belt of millions of rocks orbiting the sun, collectively known as the asteroid belt. Some of the rocks are solid, while others are thought to be floating piles of loose rubble, made of pieces ranging in size from grains of sand to mile-wide boulders. The rubble piles have an odd sort of integrity; they are held together by their mutual gravity, unlike the solid rocks we're used to, which are held together by chemical bonds.

The asteroid belt is an enigma in several other ways. For one thing, it seems much sparser than it should be. All the material in the belt today amounts to only about one-twentieth of our moon's mass, although at one time, it should have contained enough mass to form several planets as large as the Earth. Yet there's no hint of that mass today. Where did it all go?

And here's a related puzzle. For over a century, astronomers have been aware of mysterious gaps in the belt, circular gouges where no asteroids are found, like the gaps between songs on an old vinyl record. They were discovered in 1857 by Daniel Kirkwood, a former schoolteacher who learned algebra by studying a textbook with one of his pupils, and who later went on to become a math professor at Indiana University. By poring over data that astronomers had compiled, he noticed that the gaps weren't equally spaced, nor did their locations follow any obvious rules.

An important clue came in 1866 when Kirkwood rephrased the puzzle as a question about times, not distances. How long would it take, he wondered, for a hypothetical asteroid in one of the gaps to orbit the sun? By invoking Kepler's third law (a mathematical relationship between a celestial body's distance from the sun and the time required for its orbit), he was able to calculate the orbital periods associated with each gap. For example, an asteroid in the biggest gap would take about 4 years to orbit the sun: an interesting number, because it was exactly one-third as long as Jupiter's orbital period of about 12 years. Likewise, an asteroid in another of the gaps would go around the sun 5 times in the same time that Jupiter makes 2 orbits. In fact, all the gaps obeyed the same beautiful rule: Their orbital periods were always related to Jupiter's by a ratio of small whole numbers, such as 3:1, 5:2, 7:3, or 2:1.

This numerology was not coincidental. These gaps, now known as Kirkwood gaps, are the telltale sign of astronomical sync. They suggest that Jupiter's gravity is the culprit: It "resonates" with any asteroid that happens to blunder into the gaps, systematically perturbing it and eventually hurling it out of the belt.

Here's how the resonance mechanism works. Consider an asteroid with a period of about 4 years, orbiting the sun 3 times faster than Jupiter, corresponding to the 3:1 Kirkwood gap. As Jupiter makes its stately journey about the sun, following an almost circular orbit, the asteroid starts on Jupiter's shoulder and then dives toward the sun on an elongated, elliptical orbit. The sun's intense gravity whips the asteroid around like a bolo, and sends it screaming back toward Jupiter so fast that it ends up making 3 revolutions around the sun in the same time that Jupiter goes around once. At the end of its third lap, the asteroid finds itself right back where it started, hugging Jupiter's shoulder. In other words, this point of closest approach always occurs at the same place in both of their orbits.

These close encounters have a profoundly disturbing effect on the asteroid, because Jupiter is enormous, and its gravitational pull on the asteroid is most pronounced when they are closest together. Furthermore, the same disturbing effects add up relentlessly because the interaction always occurs at the same point in the orbit. Over hundreds of cycles, the periodic tugs accumulate so much that they distort the asteroid's path and cause it to become chaotic, which greatly increases its odds of leaving the belt. (By contrast, if the asteroid were not in 3:1 resonance, it would come closest to Jupiter at randomly scattered points in their orbits, so the overall effects would average out in the long run.)

Computer simulations show that asteroids flung from the belt tend to crash into the sun or fly out of the solar system. Occasionally, however, they collide with one of the inner planets. If that inner planet happens to be Earth, and if the asteroid is bigger than Mount Everest (as the dinosaur killer apparently was, based on the size of its impact crater buried beneath the Yucatán peninsula), then one can begin to see how astronomical sync could be important to us.

This argument doesn't quite answer the first riddle, however. The Kirkwood gaps are too narrow to account for all the mass that seems to be missing from the belt, making it extremely unlikely that Jupiter alone could have ejected it all. The astronomers John Chambers and George Wetherill have recently suggested an alternative solution. They propose that in the infancy of the solar system, several planetary embryos—some as large as Mars—coalesced out of the rocks in the asteroid belt (just as they did elsewhere to form the planets we see today). These proto-planets would have agitated the other rocks in the belt, nudging them into the resonant escape hatches, thereby thinning the belt more rapidly than Jupiter would have alone. Over time, some or all of these embryonic planets would themselves have set foot in the gaps, only to be ejected from the belt and never seen again.

Taking this speculation a bit further, the astronomers Alessandro Morbidelli and Jonathan Lunine suggest that one of these wayward planetary embryos may have crashed into the young Earth, flooding it with enough water to account for the oceans. It has always been a mystery to explain where Earth's water came from. The other inner planets have none, or very little. Given our position in the solar system, we seem to have much more water than we should have.

The traditional explanation is that comets, which contain a greater proportion of water than all other known celestial objects, bombarded Earth late in its formation and deposited the water we see today in the oceans, lakes, and rivers. But astronomers have begun to question that view, because the chemical composition of the water in comets is usually quite different from that seen on Earth. (Comets contain a higher percentage of heavy water, an extremely rare variant in which hydrogen, with a sole proton in its nucleus, is replaced by deuterium, with one proton and one neutron.) On the other hand, the water found in carbon-rich meteorites, believed to be fragments of asteroids, is a much closer match to that in the oceans.

The new hypothesis, then, is that our inordinate amount of water may have been the luck of the draw, the happy result of a chance collision with an icy impactor launched from the asteroid belt. If this idea turns out to be right, we have to thank astronomical synchrony not only for killing the dinosaurs and making room for our ancestors, but also for providing the water that made life on Earth possible.

As grand as sync may be at the largest scales of the cosmos, it is perhaps even more stunning at the smallest ones. Here, deep in the heart of matter, the oscillators are now electrons, the fireflies of the microworld. But unlike fireflies, which we pretended were identical for mathematical convenience, these quantum particles are thought to be truly identical. Every electron in the universe is indistinguishable from every other. They never age. They never break or chip. And their perfection makes them capable of group behavior beyond anything we've ever experienced.

In our daily lives, we are accustomed to electricity only in its chaotic form, a panic of independent particles that don't cooperate. The electrical current that powers a toaster is a mad rush of electrons, scrambling through the filament and heating it with their fury. But take the same electrons and coordinate them, and you have one of the most remarkable phenomena known to science, trillions of electrons marching in lockstep, encountering no electrical resistance, gliding through a metal without wasting any energy in the form of friction or heat. This unimaginably slippery form of electrical conduction is known today as superconductivity. Like the sympathy of clocks, it was discovered serendipitously: in this case, by asking what happens to electricity at temperatures close to absolute zero.

· Five ·

QUANTUM CHORUSES

WHEN I WAS SIX YEARS OLD, MY parents gave me a big battery to play with, the kind used in powerful camping flashlights. For some reason I had the idea to wire the two terminals together. As I walked over to my friend Casey's house to show him my new toy, I could feel the wire (and the battery) getting hotter and hotter in my hands. Electricity was circulating endlessly through the unintentional circuit I had made, and a lot of heat was being generated by the circuit's resistance to that flow of current.

At a microscopic level, trillions of electrons were banging around inside the wire, bouncing off its lattice of copper atoms in random directions, somewhat like pinballs bouncing off the bumpers in a pinball machine. In fact, the commotion is even greater than this analogy would suggest. The copper atoms are not stationary like bumpers. They're always jiggling. The higher the ambient temperature, the more violent their agitation. So a better picture would be a mob of pinballs jostling their way through an obstacle course of vibrating bumpers. Every collision with the vibrating atomic lattice impedes the flow of electrons and causes resistance.

This model of electrical conduction was familiar to all physicists by the early 1900s. It predicts that the resistance of a metal should decrease steadily as its temperature is lowered (because less shaking of the lattice means fewer and milder collisions). When experiments confirmed that prediction, some physicists began to wonder what would happen if the temperature could be reduced all the way down to absolute zero: the lowest possible temperature, where all atomic motion ceases. One camp felt that the resistance would continue to drop in tandem with temperature and then vanish at absolute zero. Another argued that the resistance would decrease to some lower limit but never disappear completely, because of the inevitable impurities and defects in any real lattice. Those imperfections would always cause some resistance, even at absolute zero.

The issue remained moot for years because no one knew how to make anything so cold. The breakthrough came when the Dutch physicist Heike Kamerlingh-Onnes devised a way to liquefy helium, allowing him to cool objects down to −269 degrees Celsius, equivalent to 4 degrees above absolute zero. He was now in a unique position to settle the question. In 1911 he found that the expectations of both camps were wrong. When he immersed a thin tube of mercury in liquid helium and lowered the temperature, the resistance of the sample decreased gradually at first, as everyone expected. But then, at a temperature of about 4.2 degrees above absolute zero, the resistance of the mercury abruptly disappeared. It didn't ramp down to zero. It plummeted. At one temperature the mercury showed a measurable resistance; drop it a tiny fraction of a degree colder, and the resistance was gone.

Kamerlingh-Onnes had just discovered superconductivity.

From the perspective of classical physics, superconductivity seems impossible. A material that conducts electricity without resistance sounds a lot like the crackpot concept of a perpetual motion machine, a machine that runs forever without suffering any friction or requiring any energy. But Kamerlingh-Onnes's observations did not violate the laws of thermodynamics; the catch is that his system was not actually functioning as a machine, in the sense that it was not performing any work on its surroundings. Still, except for that crucial caveat, superconductors do seem to be capable of a kind of perpetual motion. Later experiments have demonstrated that a pulse of electrical current can circulate around a loop of superconducting wire for years without losing any energy. As

far as we know, and implausible as it sounds, the resistance in the superconducting state is not merely close to zero; it is exactly zero. There's no way to
prove that experimentally—it would require letting the pulse swirl forever—
but such experiments do place a firm upper bound on the resistance: It's at least
a billion billion times smaller than the resistance of copper at room temperature. That's a factor less than 0.000000000000000001.

For decades after Kamerlingh-Onnes's discovery, physicists were mystified
by superconductivity. Why did the resistance drop so abruptly? And how could
it disappear at a temperature above absolute zero, when the atomic lattice would
still be vibrating? It seemed absurd to picture trillions of pinballs rushing past
the quivering bumpers without even glancing them. Something seemed to be
terribly wrong with the traditional model.

In the early 1900s, similar breakdowns were occurring throughout physics,
whenever scientists probed deep inside the heart of matter, in the microscopic
realm of atoms and electrons. For example, classical physics could make no
sense of the stability of electrons orbiting around the nuclei of atoms. The prevailing theories said that electrons should continually radiate some of their
energy away as they orbited, which would cause them to nose-dive into the
nucleus. A bad thing—and fortunately not observed.

Over the next few decades, the paradoxes were resolved, one after another,
by the creators of quantum mechanics, the revolutionary branch of physics that
proposed that matter and energy are fundamentally discrete. Max Planck
assumed that energy was packaged in tiny lumps, and found that he could then
explain the characteristic patterns of radiation emitted by materials heated to
red-hot temperatures. Albert Einstein postulated quanta of light—particles now
called photons—to explain a baffling phenomenon called the photoelectric
effect, in which light falling upon certain metals was found to stimulate the
emission of electrons. Until Einstein's work (which later won him a Nobel
Prize), no one could understand why some colors of light ejected electrons at
high speeds, while others were completely ineffectual. Niels Bohr solved the
puzzle of nose-diving electrons by sheer fiat. He declared that electrons were

confined to a discrete set of circular orbits whose angular momentum was quantized in units of a smallest denomination, a penny of angular momentum called Planck's constant. From that he was able to calculate the spectral lines—the bar code of colored light waves—that hydrogen atoms emit when excited, in convincing agreement with measurements that had gone unexplained for decades.

Later concepts in quantum theory seemed even more counterintuitive. Light was sometimes a particle, sometimes a wave. The same was true of electrons, and atoms, and all quantum objects. Even the emptiness of empty space was no longer what it seemed. In quantum field theory the vacuum became a roiling frenzy of particles and antiparticles, suddenly being born out of nothingness and then disappearing just as fast.

If one had to sum up the quintessence of quantum weirdness in a single statement, however, that statement would have to be Werner Heisenberg's famous uncertainty principle, a refined version of the adage that you can't have it both ways. The uncertainty principle expresses a seesaw relationship between the fluctuations of certain pairs of variables, such as an electron's position and its speed. Anything that lowers the uncertainty of one must necessarily raise the uncertainty of the other; you can't push both down at the same time. For example, the more tightly you confine an electron, the more wildly it thrashes. By lowering the position end of the seesaw, you force the velocity end to lift up. On the other hand, if you try to constrain the electron's velocity instead, its position becomes fuzzier and fuzzier; the electron can turn up almost anywhere.

For many years, scientists comforted themselves with the belief that these outlandish effects were limited to the subatomic domain. Today we know better. Today we understand superconductivity to be an intrusion of quantum mechanics into our everyday, macroscopic world. It gives a hint of the strangeness locked in the cellar, creeping up the stairs.

The key to the puzzle of superconductivity turned out to be the remarkable ability of electrons to pair up and move in sync. To understand how such electronic cooperation could be possible, we first need to know a little more about the rules of quantum group behavior.

All quantum particles can be classified as either "fermions" or "bosons." Fermions are territorial hermits: No two can ever occupy the same quantum state simultaneously. This rule, known as the Pauli exclusion principle, accounts for the orderly way that electrons fill the orbital shells around atoms, waiting their turn, one at a time, like polite people taking their seats in the same row of a theater. Fermions' tendency to avoid one another ultimately yields the basic laws of chemistry, most notably the structure of the periodic table, the rules for chemical bonding between atoms, and the behavior of magnets.

Bosons have the opposite kind of personality. They're gregarious. There's no limit to how many can occupy the same quantum state simultaneously. In fact, they prefer crowds: The more populated a state is, the more attractive it becomes to others. Specifically, the probability of a boson adopting a particular state is proportional to the number already in it, plus one. This means, for example, that a quantum state containing 99 bosons is 100 times more appealing than an empty one. In that sense, bosons are inveterate joiners, conformists. They love to sing along.

The first person to conceive of such a quantum chorus was Albert Einstein. The year was 1924. He had recently received a letter from a young, unknown Indian physicist named Satyendranath Bose, who had an iconoclastic idea he wanted to publish; unfortunately, his paper had already been rejected by one scholarly journal, and now he hoped to win Einstein's endorsement before trying again. Unlike the crank mail he so often received, this letter intrigued Einstein. Bose had found an ingenious way to rederive the law of radiation that Max Planck had originally worked out in 1900, the theoretical breakthrough that had started the quantum revolution. Planck's old argument had an ad hoc character—even Planck himself was not satisfied by it—but now Bose had seemingly managed to reformulate it more gracefully. Upon closer scrutiny, however, Einstein noticed the peculiar logic implicit in Bose's calculation: In the course of enumerating all the different ways that indistinguishable quantum particles could occupy energy levels, Bose had assumed new rules for counting.

The issue was somewhat like asking, How many different ways are there for two identical twins, Peter and Paul, to sit in two chairs? With normal counting, we'd say there are two ways: Paul could sit on the left and Peter on the right, or vice versa. But suppose Peter and Paul are truly identical, so that if you turned your back for a moment, you'd never know if they'd switched chairs. Then, since there's no way to tell them apart, there's essentially only one configuration: a twin in each chair. When objects are indistinguishable, said Bose, we need to count differently. Actually, Bose confessed years later that he was unaware of the novelty in his approach. His intuitive shot in the dark seemed natural to him.

Einstein extended Bose's work by considering the group behavior of *any* collection of quantum particles that obeyed these peculiar statistics. Whereas Bose had restricted his attention to pure radiation (which, like all forms of light, is made of photons that behave as if they had no mass), Einstein generalized the theory to matter (composed of particles with mass, like atoms). His mathematics predicted something astounding: When chilled to sufficiently low temperatures, such bosons (as they are now called) could display a kind of quantum sympathy. They would all act as one. Literally. The particles would lose their identities and fuse into something indescribable. Not a solid or a liquid—a new kind of matter.

Einstein's reasoning is too technical to describe here, even in metaphorical terms. But we can reach his conclusion more easily by applying the uncertainty principle that Heisenberg discovered three years later, in 1927. Although anachronistic, the following simplified argument is how most physicists today understand the phenomenon that Einstein predicted.

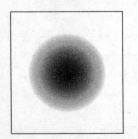

Remember, we're trying to show that an enormous number of bosons can fuse into a single entity at low enough temperatures. When you think of a boson, don't think of a point; instead, you should picture a blur, a smeared-out cloud of probability that tells you where the boson is likeliest to be found.

It might help to remember the character

named Pigpen in the old *Peanuts* comic strip. You rarely saw Pigpen; all you saw was the cloud of dust surrounding him, and you knew he was somewhere inside. Likewise, a boson is enshrouded by a spherical haze, a series of concentric shells of probability, the dark center of which is the likeliest place to find the particle itself. This center is the region of highest probability—the place where the boson "is," in our usual, pre-quantum way of thinking—although there's always a chance of finding it far out on the edge of the cloud as well.

Now imagine a flock of these clouds, all darting about at random in three-dimensional space. This flock represents a gas of bosons. The question is, What happens to this gas as we cool it down to temperatures close to absolute zero? According to Heisenberg's uncertainty principle, something very strange is bound to happen: The blurs will become even blurrier. The probability clouds will expand and thin out, meaning that the bosons can wander more widely. To see why, remember the seesaw. Chilling the bosons slows them down until they're hardly moving, which has the effect of squeezing their velocities toward a definite value (they can't go any slower than zero). Now since the velocity end of the seesaw is being pushed down, the position end must rise up; as the bosons' velocities become more definite, their positions must become less definite. In other words, they become even blurrier. Their probability clouds stretch out.

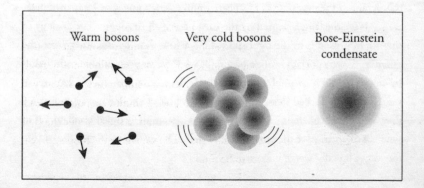

Warm bosons   Very cold bosons   Bose-Einstein condensate

At a critical temperature, the clouds broaden so much that they start to overlap, and the bosons start to mingle. As soon as that happens, said Einstein, a large proportion of them should spontaneously collapse into the same quantum state, the state of lowest possible energy. Even Einstein himself was not sure what to make of this prediction. "The theory is pretty," he wrote to his friend Paul Ehrenfest in December 1924, "but is there also some truth to it?"

Seventy-one years after its mathematical conception, Einstein's brainchild was born in a laboratory in Boulder, Colorado, in 1995. Using magnetic fields, evaporative cooling, and lasers like those in compact disc players, Eric Cornell and Carl Wieman chilled a dilute gas of rubidium atoms to less than a millionth of a degree above absolute zero, a temperature that brings gasps from even professional low-temperature physicists. Under these extreme conditions—quite possibly achieved nowhere else in the history of the universe—they observed thousands of atoms behaving as one. In 2001, Cornell, Wieman, and Wolfgang Ketterle of MIT shared the Nobel Prize in Physics for their creation of this exotic state of matter, now known as a Bose-Einstein condensate. As the Royal Swedish Academy of Sciences wrote in a press release, they had succeeded in making atoms "sing in unison."

The phenomenon of Bose-Einstein condensation is almost unimaginably alien. No one quite knows how to describe what it means. It's often said that the individual atoms coalesce into a single, giant "superatom." Others have characterized the new state as a "smeared-out, overlapping stew." My own preference is for the language used by the Royal Swedish Academy. The analogy to singing in unison is in the right spirit. Like a sine wave or any other wave, the quantum wave associated with a boson (or what we've been calling its probability cloud) has both an amplitude and a phase. In a Bose-Einstein condensate, all these waves are locked in step. Their troughs and crests line up; physicists say they are "phase coherent." Similarly, when a system of coupled oscillators is in sync, all of them have the same phase as well. The difference is that the oscillators don't literally merge into one.

Quantum phase coherence is more than an esoteric curiosity. It has given us the laser, one of the most important inventions of the twentieth century. Precisely because photons are indistinguishable particles governed by Bose-Einstein statistics, it's possible to put colossal numbers of them in the same quantum state, so that they act like a single, gigantic wave of light. The laser action is initiated when a source of energy, such as an electric current or a flash lamp, excites atoms out of their lowest energy state and pumps some of their electrons up to high energy levels (remember the watermelons being lifted onto their stools). When those atoms relax, they shed their excess energy as photons, which fly off in random directions inside the laser's cavity. Most of the photons are absorbed by the walls, but the ones that move along the line between the two mirrors at either end will continue to ricochet back and forth, reinforcing one another and inviting other photons to join their quantum state. With typical bosonic friendliness, each rebounding photon recruits new ones into the wave, through the chain reaction process known as stimulated emission: They provoke the release of other photons in sync with themselves, which amplifies the wave further, which stimulates further emission, and so on. When the wave becomes strong enough, some of it punches through the mirror at the front end (which is only partially reflective) and streaks out as an intense, narrow beam of synchronized light—a laser beam.

Quantum sync also explains how superconductivity works. The argument is tricky, because the herd behavior that we've been discussing doesn't come easily to electrons. Being fermions, they are not naturally sociable. Instead, superconductivity relies on a subtle mechanism that prods the electrons to join in pairs, at which point they become bosons and lose all inhibition. These paired electrons spontaneously form a Bose-Einstein condensate, a synchronized ensemble that encounters no resistance as it carries electrical current through a metal.

This explanation was long in coming. It required more than fifty years of insights into quantum theory, and was proposed in 1957 by the physicists John Bardeen, Leon Cooper, and Robert Schrieffer. Its most surprising innovation is the idea that electrons can form pairs. Normally we would expect electrons to repel each other, since they are all negatively charged.

The pairing mechanism is indirect. The interaction between the electrons is mediated by the lattice of positively charged ions. (Earlier, we referred to these ions as atoms. But since they are freely sharing some of their conduction electrons, they are positively charged and so should be called ions. Their positive charge is the key to the pairing mechanism.) When an electron moves through the lattice, it pulls the lattice toward it slightly, because of its opposite charge. That deformation creates a region of space with a tiny excess of positive charge, which tends to attract a second electron toward it. In that indirect sense, the two electrons are linked.

There are several ways to visualize this mechanism, none quite right, but all illuminating nonetheless. Imagine a bowling ball rolling on a waterbed. It creates a depression that tends to attract another bowling ball to follow in its tracks. Here the bowling balls are like the electrons, and the deformable waterbed is like the lattice. Or think about the drafting effect used by bicyclists in a race. The lead cyclist cuts the air, and the lowered pressure behind him pulls a second rider along in his wake. The problem with this image is that the paired electrons in a superconductor are actually quite far apart; the second one does not trail right behind the first. In that respect the paired electrons are more like a teenage couple dancing with each other at long distance, moving in step while staying at opposite ends of the floor. Although there may be many other teenagers dancing between them, there's no doubt about their being paired. After all, they're dancing together; as a physicist would say about paired electrons, their motions are "strongly correlated."

The importance of pairing is that it alters the electrons' willingness to fraternize. A single electron is a fermion, a stubborn recluse. But two electrons, once successfully paired, become effectively bosonic. (This follows from quantum theory, which shows that the distinction between fermions and bosons is akin to that between odd and even numbers; pairing two fermions makes a boson, in the same way that adding two odd numbers makes an even one.)

Once the electrons have coupled up in these so-called Cooper pairs, they become desperate to socialize with other bosons, so much so that they all pile into the same quantum state, the state of lowest energy. Then they all lose their identities and coalesce into a Bose-Einstein condensate; in the metaphor of the teenage dance party, the whole crowd is now synchronized in a collective line dance.

The Bardeen-Cooper-Schrieffer theory neatly solved a number of puzzles about superconductivity. Most important, it explained why the electrical resistance drops to zero below a critical temperature. The explanation has to do with the communal behavior of the Cooper pairs. In response to an electric field, the paired electrons march through the superconductor in rigid lockstep. Any collision with an impurity or a vibrating ion—any event that could possibly cause resistance—would have to knock a pair out of the herd and into another quantum state. But remember that the probability of joining a particular state is proportional to $n + 1$, where n is the number of bosons already in that state. The herd is billions of times more attractive than any alternative, so no pair is likely to break ranks on its own. The only way to create resistance would be to scatter billions of pairs simultaneously, an event so exceedingly unlikely as to be virtually impossible. Consequently, the resistance of a superconductor is zero, or at least smaller than anything scientists can measure.

The theory also showed that superconductivity is not a mere extension of ordinary conductivity. Previously it had always seemed paradoxical that the best normal conductors, copper and silver, are feeble superconductors; they do not superconduct even when the temperature is a thousandth of a degree above absolute zero. Seen in the light of the new theory, however, that finding began to make sense. Good conductors are good precisely because their conduction electrons ignore the lattice. But by encouraging the electrons and the lattice to go their separate ways, these materials never give Cooper pairs a chance to form. Remember, the pairing mechanism relies crucially on an electron's ability to deform the lattice (like the bowling ball rolling on the waterbed) so that a second one can follow in its tracks. If the waterbed is so stiff that the first bowling ball can't make a groove in it, there's no chance that a second one will follow. So good conductors are lousy superconductors, because they can't form the necessary Cooper pairs.

Finally, the theory explained why the resistance drops so abruptly at a certain temperature. It's much the same reason that water freezes suddenly at 0 degrees Celsius. Both processes are phase transitions, victories of self-organization over random jittering. At the freezing point, water molecules calm down just enough to allow their attractive forces to bond them into a crystal. Similarly, at the superconducting transition temperature, the atomic lattice calms down just enough to allow electrons to form Cooper pairs and coalesce into a Bose-Einstein condensate. In both cases, a fraction of a degree drop in temperature makes all the difference.

A qualitative implication of this theory was that no material should be able to superconduct at too high a temperature—perhaps 20 to 50 degrees above absolute zero—because the lattice vibrations would be too violent. And for many years that appeared to be yet another successful prediction. By trying out various combinations of metals, experimenters gradually nudged the world record up a few tenths of a degree at a time, finally grinding to a halt at 23 degrees. The insurmountable ceiling was right where it was supposed to be—at least until the mid-1980s.

It came as a shock when high-temperature superconductivity was discovered in 1986. First came the announcement of a ceramic material that turned into a superconductor at a new record temperature of 30 degrees above absolute zero. Just two years later the world record stood at an incredible 125 degrees. As of this writing, the physical basis for high-temperature superconductivity remains an enigma. It's generally believed that Cooper pairs are still involved, perhaps mediated by magnetic interactions instead of lattice vibrations. In any case, although the Bardeen-Cooper-Schrieffer theory works beautifully at low temperatures, it cannot be the whole story.

These advances rekindled interest in the possible practical applications of superconductivity. Even in its original low-temperature form, superconductivity always offered great economic and energy-saving promise. Because wires made of superconducting material have no resistance and therefore generate no heat, they can carry extremely high currents that would cause ordinary wires to burst into flame. For the same reason, they also waste much less energy. (The Department of Energy estimates that more than 7 percent of all power generated in the United States is squandered by electrical resistance and other losses in transmission; converting the power grid to superconducting technology would cut that number in half.) Aside from the benefits in efficiency, the enormous currents can be used to drive powerful electromagnets, strong enough to lift a train off its tracks, eliminating the friction between the wheels and the rails. This is the basis for the maglev (magnetically levitated) trains now being tested in Japan. In 1997, the Japanese Minister of Transport authorized construction of the Yamanashi Maglev Test Line; two years later, the MLX01 test vehicle attained a blistering speed of 343 miles per hour. Superconducting magnets are also of interest for military applications, including propulsion systems for ships, ultrasensitive detectors of submarines and underwater mines, and electromagnetic pulse-generators for frying an enemy's power grid and electronic infrastructure.

Despite its technological potential, superconducting technology has been slow to materialize in the marketplace. One obstacle has always been the frigid temperatures needed to reach the superconducting state, requiring the use of elaborate refrigeration systems available only in research laboratories. That was one reason why the discovery of high-temperature superconductivity caused such a stir: The critical temperatures could now be reached by cooling with liquid nitrogen, which is both cheap and abundant. The more serious obstacle has become the difficulty of manufacturing strong, flexible wires out of the new materials; like other ceramics, they are brittle and tend to crack easily. It's also hard to fabricate the wires in practical lengths; they tend to lose their superconductivity because of material defects when they get too long. Moreover, the most promising form of superconducting wire is encased in silver, which makes it 20 times more costly than copper wire, although the cost will decrease as demand rises. And finally, although the technology of maglev trains has been proven, their widespread use in Europe and the United States has been blocked by political and environmental concerns.

Back in the early 1960s, nobody dreamed of such things. The ramifications of the new Bardeen-Cooper-Schrieffer theory were just being worked out in labs and universities around the world. One person looking into them was a young graduate student at Cambridge University. A small, soft-spoken Welshman with black-rimmed glasses, he was about to discover some astonishing implications of quantum sync that would ultimately make superconductivity useful in unexpected settings: from medical imaging to the promise of the world's fastest supercomputers. And the path of his own career would take some of the most unexpected turns of all.

In 1962, Brian Josephson was a 22-year-old research student at Cambridge University. His subject was experimental physics, but lately he was finding himself fascinated by theoretical ideas, especially those he was learning about in Phil Anderson's lecture course. Anderson, an expert in superconductivity and solid-state physics, was visiting Cambridge for the year while on sabbatical from his position at Bell Laboratories. It didn't take him long to notice Josephson. Having him in the course "was a disconcerting experience for a lecturer, I can assure you," said Anderson, "because everything had to be right or he would come up and explain it to me after class."

One day, Josephson showed his teacher some calculations he had made on his own. He had asked himself what would happen if he connected two superconductors with a very thin layer of oxide, just one- or two-billionths of a meter thick. The picture he had in mind looked like a sandwich: The slices of bread were the superconductors, and the meat (sliced extremely thin) was the oxide layer.

Josephson could not quite believe what his equations were telling him. They said that electrical current could flow through the oxide layer without resistance. According to classical physics, that was nonsense. Oxide is an insulator. It completely blocks the flow of electrons: It's like asking them to run through a brick wall. Yet Josephson's calculations were saying that he could turn an insulator into a superconductor, converting it from one extreme to another. Instead of a brick wall, the electrons would feel like they were running through nothing at all. Instead of infinite resistance, there would be no resistance.

Josephson's prediction was based on a quantum effect known as tunneling. A quantum particle trapped deep in a well doesn't have to climb out to escape. As if by magic, it can tunnel through the walls. It doesn't even leave a hole.

Like so much in quantum theory, tunneling contradicts our common sense about how the world works. But it becomes a bit less paradoxical when we remember that quantum particles can also act like waves. Just as the sound waves from a raucous party can leak through the walls into a neighboring apartment, a quantum wave can seep through a seemingly impenetrable barrier. The odds are small, but not zero. And if the wall is paper thin, like Josephson's oxide layer, tunneling is not just a hypothetical possibility. It really does occur. Experiments have proven it. In fact, just two years earlier, Ivar Giaever, then a graduate student at Rensselaer Polytechnic Institute in Troy, New York, had demonstrated that single electrons could tunnel from one superconductor to another through an insulating barrier, though they required the helpful push of a voltage behind them. Now Josephson's calculations were saying something even stranger: Tunneling could occur without any voltage push at all.

To get a gut feeling for how paradoxical this would be, think of the flow of electricity as being analogous to the flow of water. Just as water flows downhill, electric current flows from higher voltage to lower. Now imagine two buckets, each with a small hole in its bottom, connected by a thin hose that allows water to flow between them (analogous to two superconductors connected by a thin oxide layer). If you fill each bucket with equal amounts of water, and hang one from a hook at the top of a staircase and the other at the bottom, water will drain down from the upper bucket to the lower one. But if both buckets are hung at the same level and allowed to remain there peacefully, you would never expect to see water flowing spontaneously from one to the other. Water does not flow sideways. Yet this is exactly what Josephson's equations were predicting: a flow of electricity between two superconductors at the same voltage.

What made the sideways flow possible was that he was considering a substance completely foreign to us, a substance nothing like water—a quantum fluid, a perfectly synchronized ensemble of Cooper pairs. The liquids we're used to are chaotic jumbles, made of molecules that don't cooperate. Even the smooth water in a gentle brook is, at a microscopic level, a rabble of molecules crashing into one another, sliding past one another, tumbling, bumping, and jiggling furiously. But the fluid of Cooper pairs in a superconductor is disciplined in a way we can scarcely imagine. All the paired electrons are coherent in phase; the crests and troughs of their quantum waves superimpose perfectly. If, as Josephson assumed, the oxide layer is sufficiently thin, these waves can leak through the barrier and infect the superconductor on the other side. This coupling enables Cooper pairs to tunnel through the insulator. In other words, the equations were predicting the existence of a "tunneling supercurrent."

Because this conclusion seemed so peculiar—even for quantum theory—Josephson asked his professor to take a look at his work. Anderson was happy to oblige. "By this time I knew Josephson well enough that I would have accepted anything else he said on faith. However, he himself seemed dubious, so I spent an evening checking one of the terms that make up the current." The term in question was the tunneling supercurrent. Was it really possible that the Cooper pairs could remain intact while plowing through the insulator? It seemed much more plausible that they'd split apart into single electrons and give rise to a normal current, like what Giaever had seen in his earlier experiments, a current that met with resistance as it flowed.

Casting further doubt on the issue, Josephson's thesis adviser, Brian Pippard, had previously argued that the tunneling of Cooper pairs was so improbable as to be undetectable. Roughly speaking, it was like lightning striking twice in the same spot. The probability of a single electron tunneling through an insulator was known to be tiny, so the chance of two electrons tunneling simultaneously should equal that tiny probability *squared*—an almost infinitesimally small number. Yet Josephson's math showed that the odds were about the same for two electrons as for one. "It was some days before I was able to convince myself that I had not made an error in the calculation," he wrote years later. Further reassurance came from Pippard and Anderson, who checked his work and agreed with him. The math was right. Still, all three of them felt uneasy.

Other implications of Josephson's theory were equally unnerving. His equations predicted that the strength of the supercurrent should depend on the relative phases of the quantum waves on either side of the barrier. If the phases could somehow be driven slightly out of step in the two superconductors, the supercurrent would turn on. The larger the phase difference, the larger the supercurrent, but only up to a point. Once the waves were a quarter cycle out of step, 90 degrees apart, the supercurrent would reach its maximum size. (In general, the equations predicted that the supercurrent would be proportional to the sine function of the phase difference.) To drive the waves out of step, Josephson imagined feeding electrons into the system by connecting an external source of current to the sandwich structure. As long as this imposed current wasn't too large, the equations dictated it would be carried in the form of the hypothetical supercurrent. But apparently only a limited amount of supercurrent could be conducted in this way. Try to pass more and the additional electrons would not pair. They'd break apart spontaneously, generating resistance and creating a voltage difference between the two superconductors. Then the quantum waves on either side of the barrier would unlock from each other, with their phases drifting apart at a rate proportional to the newly developed voltage. Since the supercurrent depends on the sine of the phase difference, and the phase difference is now increasing in time, the theory was saying that a constant voltage across the sandwich would produce a nonconstant, alternating current.

That prediction also violated common sense. In an ordinary resistor, a fixed voltage would produce a steady flow of current (just as water should drain steadily from the upper bucket on the staircase down to the lower one). Yet according to Josephson's equations, the tunneling supercurrent goes nowhere; it oscillates in place at a frequency proportional to the voltage. To appreciate how outlandish this is, think what it would mean for the connected buckets. If water were replaced by Josephson's quantum fluid, it would eerily slosh back and forth through the hose between the two pails. No net flow would occur. Suppose we raise the upper bucket even higher on the stairs, to increase the pressure. There would still be no net downward flow; the fluid would merely slosh faster. This effect is now called the alternating-current Josephson effect.

Another striking thing about this effect was that the ratio of voltage to oscillation frequency was predicted to be a universal constant of nature. It would always come out the same, no matter how much current was oscillating or what type of metal was used in the superconductors. The ratio is given by Planck's constant (which measures the intensity of all quantum phenomena) divided by twice the charge on the electron (the fundamental unit of electrical charge). These numbers implied that the supercurrent should tunnel back and forth extremely rapidly: A mere thousandth of a volt across the sandwich would produce an alternating current that reverses itself 100 billion times a second. For comparison, today's fastest home computers still run about 50 times slower than that.

Josephson's predictions seemed to be verging on the absurd. Were they right? The leading solid-state theorist of the day would have none of it.

John Bardeen had already won the first of his two Nobel Prizes. In 1956, he had shared the physics prize with William Shockley and Walter Brattain for their invention of the transistor. Sixteen years later, in 1972, he would receive another Nobel, this time for his solution of the long-standing riddle of superconductivity (with Leon Cooper and Robert Schrieffer) discussed earlier.

Bardeen had read young Josephson's paper. He was sure the arguments in it were spurious. In a "Note added in proof" to an article in 1962, Bardeen dismissed Josephson's purported supercurrent, asserting that "pairing does not extend into the barrier, so that there can be no such [supercurrent]."

A face-to-face showdown between the Nobel laureate and the graduate student took place in September 1962, at a low-temperature physics conference at Queen Mary College, London. Before the lectures got started, Giaever introduced the antagonists. As he later recalled,

> I introduced Josephson to Bardeen in London, when people were milling around in a big hall. Josephson tried to explain his theory to Bardeen. But Bardeen shook his head slightly and said "I don't think so," because he had carefully thought about the problem. I stood there during the short conversation. Then Bardeen left, and Josephson was quite upset. He could not understand that Bardeen was supposed to be a famous scientist.

The chairman of the session on tunneling felt it would be good to hear from both combatants. The conference room was packed in anticipation. Bardeen sat near the back of the room. Josephson went first. He gave his prepared lecture, explaining why he thought that the tunneling of Cooper pairs would be a significant effect. Then Bardeen took the podium. When he argued that pairing could not extend into the barrier, Josephson interrupted him. The exchanges went back and forth, with Josephson answering every objection to his new ideas. The mood was civil throughout, both men being calm and rational by nature. Yet Josephson seemed to be suggesting he understood the theory of superconductivity better than its creator did.

Afterward, there was hardly any discussion with the audience. Few felt confident enough to take sides. Though one person in the room, a prominent physicist from Stanford, did come to a clear conclusion about something else: He left the hall thinking his university should hire Josephson.

Meanwhile, Anderson's sabbatical had ended, and he had gone back to Bell Labs, feeling that he had become Josephson's "most enthusiastic evangelist." He and his colleague John Rowell, a skilled experimentalist, set out to look for the tunneling supercurrent. Within a few months they found it. Their measurements displayed the telltale signature of the direct-current Josephson effect—the sine wave dependence of the supercurrent on the phase—as well as the distinctive behavior expected of the supercurrent in a magnetic field. A few months later, other scientists confirmed the alternating-current Josephson effect. After those decisive tests, Bardeen graciously conceded that Josephson was right.

Within the next year, it also became clear that these phenomena were not limited to superconductivity. Richard Feynman, with his knack for getting to the bottom of things, found an elementary argument that showed how general the Josephson effects really are. He presented it to his sophomores at Caltech in 1962–63, at the end of a course later immortalized in the book *The Feynman Lectures on Physics*.

Feynman's argument shows that the Josephson effects will occur for any pair of phase-coherent systems coupled by any sort of weak link. Coherent means that each system is characterized by a single quantum wave. Weak means that the waves overlap slightly, but don't otherwise disturb each other. The overlap region spans the weak link and allows tunneling of particles across it, thus coupling the two systems. With those assumptions alone, Feynman rederived everything that Josephson had found. If the particles on the two sides of the link differ in their average energy, he predicted that they would oscillate back and forth at a frequency given by the energy difference divided by Planck's constant. This prediction was untested for years (except in superconductors) because of technical difficulties in performing the measurements. In 1997, after three decades of effort, the Josephson effect was finally seen in another phase-coherent system: superfluid helium.

Superfluid helium is a realization of the hypothetical quantum liquid that we imagined when performing the thought experiment with the buckets on the staircase. Its behavior is almost surreal. It creeps out of its containers and can flow through infinitesimal pores. It has no viscosity, so it's incredibly slippery. For example, suppose you slowly spin a bowl full of it. The container rotates but the helium doesn't. Now scoop out a cupful of the superfluid and hold it upright, an inch over the bowl. Defying gravity, a solitary drop of fluid climbs up the inside wall of the cup, runs over the lip, and rains back down into the bowl. As soon as it falls, another drop starts climbing. Like something out of science fiction, the superfluid pours itself back into the bowl, one drop at a time, until the cup is empty.

This weird behavior is a manifestation of quantum sync. All liquids become highly ordered when cooled to very low temperatures. Normally they freeze into a crystal. But the two isotopes of helium, helium-3 and helium-4, never solidify, at least not at ordinary pressures. They remain liquids all the way down to absolute zero. The liquid resolves the paradox by ordering itself in a different sense: It undergoes Bose-Einstein condensation and becomes a quantum chorus. Here, the bosons are the helium-4 atoms (or pairs of helium-3 atoms, analogous to Cooper pairs). At extremely low temperatures all the atoms slow down, which causes their quantum waves to stretch out, by the Heisenberg argument mentioned earlier. At a critical temperature the waves overlap and spontaneously fall into the same quantum state, synchronizing trillions of atoms into a phase-coherent superfluid.

In 1997 a team of physicists at the University of California at Berkeley, led by Seamus Davis and Richard Packard, turned the thought experiment with the buckets into reality. They took two tiny pools of superfluid at different pressures and coupled them by a weak link: an ultrathin, flexible membrane perforated by thousands of narrow pores. According to Feynman's analysis, the superfluid should oscillate back and forth through the pores at a frequency proportional to the pressure difference (whereas a normal fluid would simply flow from the high pressure side to the low pressure side). The experiments are extremely difficult, partly because helium is not charged, which means its flow cannot be detected as an electrical current, and partly because the pores must be made extraordinarily small, about a hundred times smaller than a bacterium.

Davis and Packard had already spent a decade searching in vain for the predicted oscillations. Now they had a new strategy, and a new team of graduate students ready to try it. Their plan was to deflect the membrane momentarily, squeezing the fluid on one side and creating a transient difference in pressure. Then, as the membrane relaxed to equilibrium, they would monitor the vibrations induced in it by the oscillating superfluid. The signature of the alternating-current Josephson effect would be an oscillation of decreasing frequency, a whistle that dropped in pitch as the pressure difference returned to zero. But even with the finest oscilloscopes, the graduate students hadn't managed to find anything remotely like that. They blamed it on too much noise in the system. After months of trying, they were dejected and ready to give up.

Their adviser Packard told them to turn off the oscilloscope, get some headphones, and *listen* for the vibrations. The students said no, it won't work, there's nothing there. "They really didn't want to do it—in the end they simply argued that they couldn't do it because they didn't have any headphones in the lab," Packard recalled. So he went to a nearby electronics shop and bought the headphones for $1.50. The students said the connector was wrong. Packard went back and bought an adapter.

Reluctantly, grad student Sergey Pereverzev plugged in the headphones and flipped a switch to start the experiment. He almost fell off his chair. His ear immediately detected what the oscilloscope had missed: a high-pitched whistle that gradually dropped in tone, like the sound of a falling bomb. Exactly what the theory predicted.

Over the past 40 years, a number of practical applications have been found for these remarkable manifestations of quantum sync. Josephson's superconducting sandwiches, now known universally as "Josephson junctions," have spawned the most sensitive detectors known to science. For instance, a device called a SQUID (for superconducting quantum interference device) takes advantage of the extreme sensitivity of a supercurrent to a magnetic field. A SQUID can measure a displacement a thousand times smaller than an atomic nucleus, or a magnetic field 100 billion times weaker than Earth's. SQUIDs are used in astronomy, to detect faint radiation from distant galaxies; in nondestructive testing, to spot hidden corrosion beneath the aluminum skin of airplanes; and in geophysics, to help locate sources of oil deep underground.

A SQUID consists of two Josephson junctions connected in parallel by a loop of superconducting material. (To picture this, hold your arms above your head and clasp your hands together. Your two elbows are the two Josephson junctions, and the circle formed by your arms and shoulders is the superconducting loop.) The principle underlying a SQUID is that variations in a magnetic field alter the phase difference between the quantum waves on either side of its two junctions, and therefore change the supercurrents tunneling through them. Just as ripples on a pond can either add up when they collide (if a crest meets a crest) or cancel each other out (if a crest meets a trough), the quantum waves in the two arms of a SQUID interfere in a way that depends sensitively on their phases, and hence on the amount of magnetic flux passing through the loop. In this way, a SQUID transforms tiny variations in magnetic flux into measurable changes in current and voltage across the device, allowing ultrafaint electromagnetic signals to be detected and quantified.

Some of the most dramatic applications are in medical imaging. With an array of hundreds of SQUID sensors, doctors can pinpoint the sites of brain tumors and the anomalous electrical pathways associated with cardiac arrhythmias and epileptic foci (the localized sources of some types of seizures). The SQUID array maps the subtle spatial variations in the magnetic field produced by the body. The resulting contour map enables computers to reconstruct the region inside the tissue that produced the signals. These procedures are entirely noninvasive, unlike conventional exploratory surgery. Although the high price of the multichannel imaging machines has kept them from gaining widespread acceptance, in the long run they have the potential to reduce health-care costs substantially. For example, localizing an epileptic focus with SQUIDs takes about three hours, whereas the alternative method of implanting electrodes on the patient's brain may last as long as a week and cost $50,000 more.

Josephson junctions have also been considered as possible components for a new generation of supercomputers. One attractive feature is their raw speed: They can be switched on and off at frequencies of several hundred billion cycles per second. But perhaps even more important, Josephson transistors produce a thousand times less heat than conventional semiconductors, which means they can be packed tighter on a chip without burning themselves up. Dense packing is always desirable because smaller computers are faster. By using less wire, they are less burdened by the speed of light, which ultimately determines the time it takes for signals to travel from one part of the circuitry to another.

Seduced by these appealing qualities, IBM famously invested 15 years and $300 million in a high-profile project to build a superconducting computer, an ultrafast, general-purpose machine whose logic and memory chips would be made out of Josephson junction switches. It was a natural idea, since some types of junctions have two stable states—one at zero voltage, another with a positive voltage. Any two-state device is a candidate for a switch, corresponding to the on-off, 0-1 binary logic that computers employ. Similarly, the absence or presence of a particular bit of memory would be encoded as the absence or presence of a voltage in the corresponding Josephson memory element.

When IBM abandoned the project in 1983, the reason cited was the difficulty in developing a high-speed memory chip. Management judged that by the time its new computer could be built, its performance would not be far enough ahead of the semiconductor competition to warrant the revolutionary change in approach. Since then, Hitachi, NEC, Fujitsu, and other Japanese companies have continued to chase the dream of a Josephson computer.

Ironically, Josephson himself played almost no part in the developments that stemmed from his work. After he received the Nobel Prize in 1973, at age 33, he quit doing mainstream physics and became preoccupied with paranormal phenomena: homeopathy, ESP, remote viewing, even psychic spoon bending. He continues to work on these questions today. His attitude is that they deserve more attention from science and should not be "blacklisted," as he feels they currently are.

My students laugh when I tell them what became of Josephson. Among my colleagues, the reaction is similar; they typically shake their heads and mutter about how he's gone off the deep end, while a few become downright angry, furious that he would lend his stature to a field populated mainly by charlatans and their gullible supporters. That hostility was on full public display recently, thanks to a flap that Josephson deliberately provoked.

On October 2, 2001, Britain's Royal Mail service issued a special set of stamps to commemorate the one-hundredth anniversary of the Nobel Prize. The stamps were accompanied by a booklet in which a British winner in each of the six prize categories—physics, chemistry, medicine, peace, literature, and economics—was invited to write a small article about his award. The physicist they happened to select was Josephson. Here's what he wrote:

PHYSICS AND THE NOBEL PRIZES
Brian Josephson, Physics Department, Cambridge University

Physicists attempt to reduce the complexity of nature to a single unifying theory, of which the most successful and universal, the quantum theory, has been associated with several Nobel prizes, for example those to Dirac and Heisenberg. Max Planck's original attempts a hundred years ago to explain the precise amount of energy radiated by hot bodies began a process of capturing in mathematical form a mysterious, elusive world containing "spooky interactions at a distance," real enough however to lead to inventions such as the laser and transistor.

Quantum theory is now being fruitfully combined with theories of information and computation. These developments may lead to an explanation of processes still not understood within conventional science such as telepathy, an area where Britain is at the forefront of research.

Telepathy? Explained someday by quantum mechanics? The reaction among physicists was fast, predictable, and allergic. "It is utter rubbish," said David Deutsch, a quantum physicist at Oxford University. "Telepathy simply does not exist. The Royal Mail has let itself be hoodwinked into supporting ideas that are complete nonsense." "I am highly skeptical," said Herbert Kroemer of the University of California at Santa Barbara, himself a Nobel laureate. "Few of us believe telepathy exists, nor do we think physics can explain it. It also seems wrong for your Royal Mail to get involved. Certainly, if the U.S. postal services did something like this, a lot of us would be very angry." The Royal Mail mustered a limp defense. "The trouble is that there are only a couple of British physics prize winners we could have asked, and we picked Josephson," said a spokesman.

The condescension of the physics community is unwarranted. Josephson was a hero, and still is. When I read his discussions of paranormal phenomena, they don't strike me as strident, or nonsensical on their face. He seems truly curious about these possibilities. He wants scientists to look into them more carefully. Quantum theory is plenty weird in itself, nearly as far-fetched as the things he is thinking about. A hundred years ago, no one would have believed that electrons could synchronize by the billions and pass through impenetrable barriers.

This is not to say I agree with Josephson. His belief that "some people can bend metal in situations where they are not in physical contact with it" is tough to swallow. In any case, when I think about what has become of him, my main feeling is one of wistfulness. Even after 30 years apart, many of us in the physics community still miss him.

Six

BRIDGES

I T WAS A TIME OF HIDDEN PARALLELS, of lives in imperceptible synchrony. The year was 1962. Brian Josephson was beginning graduate school. Arthur Winfree was entering college. Michel Siffre was shivering in a cave deep underground in France, subjecting his body to the unknown effects of "life beyond time." Norbert Wiener was riding his unicycle through the corridors of MIT, eating peanuts and smoking his cigar, on the lookout for his next audience. Lev Landau lay clinging to life in a Moscow hospital, comatose for months after his devastating car accident. All had made, or were destined to make, seminal contributions to the science of sync. Yet all were oblivious of one another. It was only decades later that we began to realize the true depth of the ties among them, and between them and Christiaan Huygens, who, almost exactly 300 years earlier, sick in his bedroom, observed his pendulum clocks swinging in sympathy. We now see their work as part of an intricate whole, bridged by mathematics.

The first bridge to be noticed joins the familiar world of everyday experience to the strange world of the quantum. In 1968, D. E. McCumber of Bell Laboratories and W. C. Stewart of RCA Laboratories independently figured out how to analyze the electrical characteristics of a Josephson junction as if it were an ordinary element in a circuit. Just as a resistor obeys Ohm's law (the current through a resistor is proportional to the voltage across it), a Josephson junction obeys its own distinctive relationship between current and voltage. Specifically, when an externally imposed current is driven through a junction, the current splits up and flows through three separate channels, each representing a different conduction mechanism. Part of the current is carried by Cooper pairs of electrons—the weird supercurrent that suffers no resistance as it tunnels through the insulating barrier—while the remaining parts are carried by normal, unpaired electrons and by displacement current (a form of conduction associated with the changing voltage across the junction).

By taking all three pathways into account, McCumber and Stewart found that the junction's dynamics were most naturally expressed in terms of its changing phase, a measure of how out of sync the quantum waves are from one side of the barrier to the other. This was already a novelty: In the usual laws of electricity, there's no vestige of anything bearing the stamp of quantum mechanics. Looking deeper, McCumber and Stewart noticed that the equation for the electrical oscillations was an old friend in disguise, an equation known to any student of freshman physics.

It was the equation for the motion of a pendulum.

This is the sort of coincidence that fills a mathematician with awe. "It is a wonderful feeling," said Einstein, "to recognize the unity of a complex of phenomena that to direct observation appear to be quite separate things." On the surface, Huygens's pendulums and Josephson's junctions seem like polar opposites. Pendulums are comfortable and familiar, human in scale, as common as a child playing on a swing, as cozy as the ticking of a grandfather clock. Superconducting junctions are alien, almost otherworldly, no bigger than a bacterium, with frenzied electrical oscillations 100 billion times faster than a heartbeat, the surreal consequence of electrons passing through impenetrable barriers like ghosts walking through walls. No matter. Those differences are gloss. Fundamentally, the dynamics of Josephson junctions and pendulums are the same. Their patterns in time are identical: two variations on a single algebraic theme.

. . .

Unfortunately, the recognition of an old friend also brings up an inescapable difficulty. The equations for the pendulum are nonlinear.

Specifically, the gravitational torque on the pendulum is a nonlinear function of its angle. You can understand why by imagining how hard it is to hold a barbell away from your body with your arm extended at various angles: straight down, sideways at shoulder level, directly overhead, and so on. (It's important here not to confuse the difference between weight and torque. Wherever the barbell happens to be, gravity pulls down on it equally hard—the downward pull is determined by its weight alone. But at some angles, gravity also tends to twist your arm, wrenching it downward. Torque measures the strength of that twisting effect.) When your arm is straight down, there's no torque at all, no tendency to twist your arm to either side. As you rotate your arm up at a slight angle—still almost straight down, but cocked a little to one side—gravity exerts a small torque. At first the torque grows nearly proportional to the angle. The torque at 2 degrees deflection is double that at 1 degree, to a very good approximation. For these small angles of deflection, the torque is said to be a linear function of the angle: double the angle, double the torque. In this case a graph of torque versus angle would fall on a straight line (hence the term *linear*).

But the approximate linearity breaks down as the angle increases. The torque grows slower than you'd expect; it falls below a straight-line extrapolation of the earlier trend. The largest torque occurs when your arm is sticking straight out from your side, at a 90-degree angle. It's tough to hold a barbell like that for long. If you lift your arm even higher, above your shoulder, now the torque begins to decrease, eventually reaching zero torque when the barbell is directly overhead. Thus, the curve of torque versus angle looks like an arch. It bows down. It's definitely not linear. In fact, it's an arc of a sine wave.

Now we see the connection to the Josephson junction. This sine function is the same one that appeared earlier in the direct-current Josephson effect, where the supercurrent is proportional to the sine of the phase across the junction. That's the analogy: The phase across the junction is like the angle of the pendulum. As it turns out, all the other terms in the equation have counterparts as well. The flow of normal electrons corresponds to the damping of the pendulum caused by friction. The pendulum's mass is like the junction's capacitance. And the torque applied to the pendulum is like the external current driving the junction.

Such mechanical analogies are always valuable in science. They make the unfamiliar familiar. Here the analogy allows us to transfer our intuition about pendulums to Josephson junctions. For example, when the junction is in steady operation, the phase is constant. In that case, there are no dynamics, and nothing to study; the junction acts like a perfect superconductor, with only supercurrent flowing across it. The mechanical analog would be a pendulum twisted to the side by a constant torque, resting motionless, cocked at an angle below the horizontal. Friction and inertia are absent, since nothing's moving. Gravity alone balances the applied torque. This simple case occurs only if we send less than a critical amount of direct current through the junction.

The more interesting case is when we drive the junction with more than the critical current. Then the phase suddenly begins to change in a complicated way as a function of time. Once the phase starts varying, a voltage develops across the junction. Then, because of the alternating-current Josephson effect, a supercurrent starts to oscillate back and forth between the superconductors. Meanwhile, this voltage also drives some ordinary, unpaired electrons through the resistive channel, while the displacement current vies for its share of the total current as well. So all three channels become active. Their interplay produces a bewildering ebb and flow of current among the three of them. All of this complexity can be traced to the nonlinear dynamics of the phase across the junction. In mechanical terms, you should picture a pendulum rotating over the top at variable speed, hesitating on the upswing, accelerating on the downswing, all the while balancing the applied torque against the fluctuating combinations of friction, gravity, and inertia.

If we make things even more complicated and allow the torque itself to vary in time, like the back and forth agitation of a washing machine, the pendulum's whirling can become chaotic, rotating this way and that, changing direction haphazardly. The verification of the corresponding electrical spasms in a Josephson junction was one of the early experimental triumphs of chaos theory. Before that, physicists had always seen the pendulum as a symbol of clockwork regularity. Suddenly it was a paradigm of chaos.

The essential point is that the dynamics of a whirling pendulum and a Josephson junction are governed by the same equation, and that equation is nonlinear. As stressed earlier, nonlinear problems are rich, fascinating, and very hard. They lie at the frontier of mathematics, and far beyond. The advances in chaos theory in the 1970s and 1980s (dealt with in greater detail in the next chapter) opened our eyes to the dynamics of a driven pendulum or Josephson junction, and allowed us to decipher them.

The connection between pendulums and Josephson junctions is just one of many remarkable bridges in the landscape of sync. My colleagues and I recently stumbled across another one, perhaps even more unexpected, linking populations of biological oscillators to the dynamics of Josephson junctions coupled together in large arrays. The meaning of this latest connection remains cryptic, but it seems likely to be important, because it joins two great bodies of science. One part deals with the ancient observations of life in sync: the firefly trees of Thailand and Malaysia, the nightly choruses of crickets, the daily cycles of plants and animals entrained by the sun. The other deals with the study of inanimate sync, beginning with Huygens and his sympathetic pendulum clocks, a line that fell dormant for hundreds of years, only to be reawakened with the invention of the marvelous oscillators of the twentieth century: electrical generators and phase-locked loops, lasers and transistors, and now superconducting Josephson junctions. Although it was always clear that groups of living and nonliving oscillators were each prone to synchronize spontaneously, it was only in 1996 that we realized how similar the underlying mechanisms can be. The resemblance, it turns out, is familial—a sign of the same mathematical blood.

The connection was uncovered through the study of Josephson junction arrays, an architecture that corresponds to the next level in the hierarchy of sync. We have already discussed the lowest, subatomic level, the one considered by Josephson himself—trillions of synchronized Cooper pairs of electrons, tunneling back and forth coherently through a junction, creating the supercurrent that oscillates across its insulating barrier. The next step is to couple many of these electronic oscillators together into an array and explore the synchronization among them. In terms of an earlier analogy, the Cooper pairs are like the individual violinists in an orchestra, harmonizing to form a well-disciplined string section—a Josephson junction. Then many different sections (strings, woodwinds, percussion) blend into an even larger ensemble, an orchestra—an array of Josephson junctions. No conductor is assumed, however; the array is supposed to synchronize itself.

The challenge is to predict the group behavior of Josephson junctions, given what is known about them as individuals. The question is important because Josephson arrays are used in many modern technologies, from brain scanners and other kinds of medical imaging equipment, to detectors of electromagnetic radiation at the wavelengths of interest in radio astronomy and atmospheric pollution monitoring. The U.S. Legal Volt (the official standard of voltage that allows laboratories worldwide to compare their results) is maintained by the National Institute of Standards and Technology, using an array of 19,000 Josephson junctions coupled in series. Circuit designers would love to be able to predict the best layout for an array serving a particular function, but because of the intractability of the governing nonlinear equations, they've had to rely on instinct, or trial and error.

Theorists have tried to offer guidance by forcing the equations into a linear mold, at the cost of drastic approximations. This Procrustean approach has occasionally shed light on the most symmetrical kinds of collective behavior, such as the perfectly synchronized state where all the junctions oscillate in lockstep. But as an exploratory tool, linear theory is miserable. It's too myopic to offer any hint of the myriad alternative ways an array might organize itself.

Only nonlinear dynamics, with its emphasis on geometry and visualization and global thinking, is up to the task. Of course, the job is daunting, to look at all possibilities at once, to explore the dynamics of hundreds of nonlinear equations, corresponding to a mathematical flow in an abstract space of hundreds of dimensions. But around 1990, buoyed by the successes of chaos theory, the nonlinear community was ready for this challenge. Theorists were feeling confident and hungry. Mathematical biologists had already plunged into high-dimensional spaces, groping around in the dark, trying to understand their idealized models of coupled fireflies and neurons and heart cells. This was the new perspective that Kurt Wiesenfeld, a young physicist at the Georgia Institute of Technology, wanted to bring to the analysis of Josephson arrays.

By 1990, Kurt had already made a name for himself. In 1987, he had cowritten the paper that introduced the concept of "self-organized criticality," an ambitious theory that promised to explain why so many complex systems seem perpetually poised at the brink of catastrophe. The theory was later applied to explain the peculiar statistical patterns observed in mass extinctions, earthquakes, forest fires, and other complex processes in which domino effects propagate through the system, usually producing small cascades and occasionally cataclysmic ones. The work was bold and controversial. Most physicists saw it as an important advance in our understanding of complex systems, though some skeptics dismissed it as the latest fad. One joker referred to it as "self-aggrandizing triviality."

At the time, Kurt was a postdoctoral fellow at Brookhaven National Laboratory. Now he was an assistant professor, looking to venture out on his own. He'd always been fascinated by coupled nonlinear oscillators, and had even dabbled specifically in coupled pendulums at the beginning of his work on self-organized criticality. So he felt at home with the circuit equations for Josephson junction arrays, which reminded him of the pendulum problems he was used to. His entry to the field came when he began collaborating with Peter Hadley, a graduate student at Stanford University, and his adviser Mac Beasley, an expert on superconductivity, who had already realized that nonlinear dynamics should have something to offer to the analysis of Josephson arrays. When they enlisted Kurt's help, the project took off. It was a strong team. Hadley was the hardworking grad student, resourceful and sharp at computer simulations. Beasley was the lanky, white-haired adviser, savvy, full of aphorisms and experience. Kurt was a top gun in nonlinear dynamics, one of the best around.

They decided to focus on "series arrays," with all the junctions chained end to end. That sort of architecture was the most tractable from a mathematical perspective, and also of technological interest for applications to power generation. Although a single Josephson junction produces only about a microwatt of power—too puny to be practical for anything—its output could be greatly amplified by cooperation. Just as an audience clapping in sync makes a lot more noise than any individual person, a synchronized array of Josephson junctions would be a much more potent source of radiation than any solitary one. For example, if you could find a way to coax a thousand junctions to oscillate in phase, the power delivered to another device—a "load" in parallel with the array—would be amplified a millionfold. (The combined power is proportional to the square of the number of junctions.) The hard part is figuring out a way to sync them. No one knew the optimal architecture for the circuit or the best kind of load. In fact, no one really knew why arrays should or should not synchronize at all. This was a fundamental issue, a roadblock for the whole field.

Kurt and his collaborators knew that the electrical characteristics of the load—the way it impeded the flow of current—were likely to be crucial. (With no load at all, the junctions would never synchronize; they wouldn't even be able to feel each other's electrical oscillations.) The simplest kind of load would act like a resistor, passing current in proportion to the voltage across it. Or it might behave more like a capacitor (which blocks direct current but is permeable to alternating current) or an inductor (which has the opposite characteristics: porous to direct current, resistant to rapidly alternating current). In general, the load could involve some combination of those three kinds of impedance, weighted with different strengths—a lot to choose from.

By simulating dozens of cases on the computer, the team mapped out the stability characteristics of the synchronized state, and learned which loads best synchronized the array. But they also ran across something they weren't looking for, something eye-catching and hard to miss. When the arrays didn't sync, they usually fell into a different kind of order: All the junctions oscillated with the same period but stayed as far out of step as possible, almost as if they were repulsive. The team referred to this curious mode of organization as the antiphase state; later it came to be known as the splay state.

For two junctions, the splay state is like what Huygens observed when his clocks were in sympathy: The pendulums swing at the same rate, but exactly half a cycle out of step. One says *tick* when the other says *tock*. With more than 2 junctions, the splay state divides the cycle into equal parts. If there are 10 junctions, they will execute identical motions, splayed a tenth of a cycle apart. All move in the same way, equally staggered in time. It's tempting to visualize this group behavior as a graceful choreography, a wave rippling through the array, but that image is misleading. The wave doesn't necessarily propagate from one junction to its neighbor; they can take their cues in any order. If the electrical oscillations were mechanical instead, a splay state would look something like a row of dancing robots, all performing the same contorted sequence of moves, but arranged arbitrarily in space: One robot does something, then far down the line, another does the same thing, then back somewhere else, another starts in. All permutations are allowed. The robots can dance in any order; each ordering is a valid splay state. They differ only in spatial arrangement, not in the moves performed or the timing between them.

The larger the array, the more permutations are possible. The number grows extremely rapidly, even faster than exponential. With 5 junctions, there are 24 splay states. With 10, there are 362,880. Kurt thought this explosive proliferation might offer a basis for a promising memory architecture for a future Josephson computer. Each memory could be stored as a different splay state. Instead of a static collection of 0s and 1s, it would be encoded as a dynamic pattern, a swirling dance of electrical activity in the array. (Neuroscientists believe that our memory for odors works something like this, where the oscillators are neurons in the brain's olfactory bulb, and different patterns of excitation encode different smells.)

With only a few junctions, you could make a gigantic memory, as large as you wanted. There was only one catch: For the scheme to work, each state would have to be stable, to prevent corruption by random noise in the circuitry. So now the question became, Are the splay states stable? And how does their stability depend on the load? At the time, Kurt was unable to solve the problem mathematically. More important, he realized he still lacked a global understanding. Besides synchronized states and splay states, what else might be out there? And how does it all fit together? His goal was ambitious: to understand all possible kinds of collective behavior, for any number of junctions in series, and in parallel with any kind of load.

When I met Kurt at a conference in Texas in 1990, we felt an immediate rapport. We were about the same age, with similar backgrounds and taste in scientific problems—and we found that we laughed a lot together. Now, as he told me about his vision for the Josephson array problem, I thought it might be fun to work on it together. Kurt, perhaps feeling a little guilty about what a treat this was going to be, reminded me about the possible technological applications of the work (the sort of serious justification you're supposed to offer if someone asks you why you work on what you do). But to be honest, the applications were not the real reason we were interested in these arrays. The main attraction was pure curiosity, just the pure pleasure of working out the math for a beautiful system of coupled oscillators.

In particular, there was something beguiling about the equations themselves. Every junction appeared to be coupled equally to every other. Even though they were physically connected in series, like the links in a chain, the equations made them look like they were connected all-to-all. That surprised me, and delighted me. I was already familiar with that strange, supersymmetrical kind of connectivity from my previous work on the Peskin model of heart cells and the Winfree and Kuramoto models of biological oscillators. In those settings, all-to-all coupling was chosen purely for expedience. No one knew the right equations anyway, so it was natural to start with the easiest case. Though, of course, it was a caricature: Real heart cells and fireflies interact more strongly with their neighbors than with those far away.

So when the same old egalitarian coupling appeared in the equations for the

Josephson array, I nodded knowingly. Here comes the standard approximation. No, no, Kurt told me, it's really like that. All-to-all coupling is rigorously correct here. It comes straight from the circuit equations, a consequence of the fact that when junctions are in series, the same amount of current flows through each of them, like water passed along in a bucket brigade. He promised to send me a long letter after the conference was over, with all the details spelled out.

Even before I opened the envelope, I knew from the way he wrote my address that he was going to be fun to work with. He printed in calligraphy—graceful, undulating letters, precise and whimsical at the same time. Over many years of grading students' tests, I'd come up with a kind of amateur handwriting analysis that never failed: Whenever all the answers were printed in tight little letters, machine perfect, almost as if typewritten, I knew the student was going to be near the top of the class. This rule, by the way, says nothing about messy handwriting. A student who scrawls his answers might be muddled or brilliant or anywhere in between. But calligraphy? That had to be a good sign.

Kurt suggested we begin with the most idealized possible problem: two identical Josephson junctions connected in series, and driven by a constant current. Suppose the load is a resistor, again the most vanilla choice, and instead of the usual three channels for current flow in each junction, assume each has only two pathways, one for supercurrent and another for normal current. (For certain kinds of junctions, the third pathway—the displacement current—can be neglected, to a good approximation.)

The advantage of these simplifications was that we could then visualize the system's dynamics by drawing ordinary two-dimensional pictures. At any given instant, each junction has a well-defined phase, just as a pendulum captured in a snapshot appears cocked at some angle. Graphing one phase horizontally, and the other vertically, we could represent all possible combinations as a point in a square, with 360 degrees of possible phases on either side. This square is called the system's "state space." It has an amusing geometrical property, reminiscent of the old video games where a spaceship sailing off the right edge of the screen magically reappears on the left edge, and one crashing into the bottom reappears on the top. The state space for this Josephson array would have the same magical feature, because a phase of 360 degrees is physically indistinguishable from one of 0 degrees (just as a pendulum hanging straight down would still be hanging straight down if you rotated it by a full turn). Since the left and right edges of the square correspond to the same physical state, mathematicians imagine them as being fused seamlessly into one, as if you rolled a piece of paper into a cylinder and taped the edges together. Furthermore, the top and bottom edges are also the same, so they should be taped together too, which means the cylinder is bent around into a doughnut shape, forming a surface known as a torus.

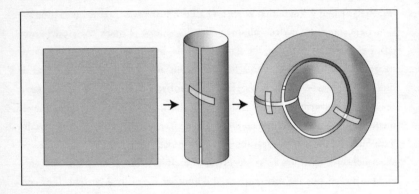

The conclusion, then, is that the state space for this simplest of Josephson arrays is equivalent to the surface of a torus. Every point on the torus corresponds to an electrical state of the array, and vice versa. As time passes and the array changes its state from moment to moment, the corresponding point on the torus glides smoothly from place to place, like a speck of dust carried on the surface of a gentle stream. The flow pattern for this imaginary stream—its whorls and eddies, backwaters and torrents—are all inherent in the circuit equations for the array. Given the present values of the phases, the equations dictate how they will change in the next instant.

The equations are nonlinear, so we couldn't hope to solve them explicitly, but we thought it might be possible to infer the overall qualitative features of the flow pattern. For example, stagnation points (places on the torus where the speck gets stuck) would correspond to states of electrical equilibrium for the array, with all currents and voltages constant in time. The stability of such states could be assessed by imagining the speck to be nudged away from them; if it always returns, as if it were being sucked down a drain, the equilibrium state is stable. Or suppose the flow pattern contains a closed loop, an eddy around which a speck can circulate endlessly, always revisiting its starting position after a certain amount of time. Such a loop would signify a periodic, repetitive form of behavior—an electrical oscillation in the array. Kurt and I knew that such loops were bound to occur, but we didn't know anything about their stability, whether they'd funnel nearby states into themselves or not.

The simplest loop is the synchronous oscillation, where the phases of both junctions are equal at all times. The corresponding trajectory flows along the main diagonal of the square. It starts in the lower left corner, then travels northeast until it exits at the top right corner, where it instantly returns to the lower left (since 360 degrees and 0 degrees correspond to the same phase). When viewed on the square, the trajectory appears to jump discontinuously from one corner to the other, but on the torus—the true state space for the system—there is no jump. The transition is seamless.

When we analyzed the overall flow pattern, we were shocked to find that every other trajectory repeats itself in a similar way. Every solution is periodic. On the face of it, that might not sound so surprising. A pendulum swinging to and fro would always repeat its behavior, at least in the simple textbook case where there's no friction in its bearings and no air resistance. In that case, it doesn't matter whether you start the pendulum swinging in a large arc or a small one—either way, it always repeats. The same is true for other kinds of "conservative" mechanical systems, hypothetical idealizations where all forms of friction and dissipation are imagined to vanish, and mechanical energy is perfectly conserved, with none lost to heat. But that's precisely why the periodic behavior of the Josephson array was such a shock to us. This array was oozing with friction. In electrical terms, friction is resistance. The junctions themselves contained an effective resistance (corresponding to the pathway for normal current), and the load was a resistor. Yet somehow this array was impersonating a conservative system.

Kurt and I wondered if this paradoxical behavior might be an artifact of studying only two junctions. With more than two, maybe the system could spread its wings and show a more representative range of behavior. I had some old computer programs lying around from my earlier work on biological oscillators, the ones I'd used to simulate the Winfree and Kuramoto models, with hundreds of colored dots running around a circular track, and also one for the Peskin model of heart cells, where it proved so helpful to strobe the system whenever one of its oscillators fired. All these programs were easy to adapt to the Josephson array equations. With Kurt now back at Georgia Tech and me back at MIT, it made sense to divide the labor. Kurt and his student Kwok Tsang pursued the mathematical analysis for more than two junctions, while I tried to get the simulations rolling.

Ten junctions seemed like a good starting point: few enough to be manageable, but too many to visualize easily. Instead of a flow on a square or the surface of a torus, the trajectories now lived in a 10-dimensional space. Undaunted, my computer programs crunched through the nonlinear equations, inching forward one tiny step at a time, and then displayed the evolving phases of the junctions as 10 dots running around a circular track. The images were dizzying. The dots swirled around, leaving an overwhelming impression of swirling but not much else. It was especially difficult to perceive any gradual adjustments in relative positioning. Some relief was provided by the strobe trick. When a preassigned junction reached a certain phase, an imaginary flash went off and illuminated the phases of the other nine junctions. That took care of the swirling, but there were still 9 dots to watch simultaneously. Following 9 dots amounts to picturing a 9-dimensional space.

The human brain cannot readily visualize more than three dimensions, and the flat screen of the computer limited the display even further to two. I needed some way to expand my mind, to try to grasp what was going on in this nine-dimensional wilderness. After playing around for a while, I finally settled on a multipanel format, like in those hokey movies from the 1960s, with split screens showing different actors in each one. One panel graphed the phase of junction number 2 versus junction number 3, with one axis dedicated to each. Other panels showed the same thing for junctions 3 versus 4; 5 versus 6; and so on. Junction number 1 was assigned to trigger the strobe: Whenever it crossed a designated starting line (a specific phase in its cycle), the computer plotted the corresponding point in each panel, representing the simultaneous phases at that instant. So the computer screen was filled with panels, continuously updated with each stroboscopic flash.

Before looking at the system through these new theoretical goggles, I needed to prepare myself for what I might see. In the worst case, if the solutions to the equations were horribly complicated, the dots would hop around wildly in each panel, gradually filling out an amorphous blob. If they contained a bit of structure, the blob might be lacy, with striations in it. Or if things were as paradoxically simple as they were for two junctions, each dot would keep landing on the same point, boring a hole into the computer screen, never budging from its birth pixel. That incessant repetition would signal that all the trajectories were still periodic (because for a periodic solution, whenever junction 1 crosses the starting line to trigger the flash, junctions 2 and 3 would always appear in their proper places, and likewise for every other panel).

I unleashed the computer and stared at the screen. After a while, a single dot appeared simultaneously in each panel, meaning that junction 1 had completed one lap and fired its strobe. Then another lap, and another. In every panel, the dots kept landing close to the original ones, but not quite on top of them. That was already interesting. The near misses meant the trajectories for 10 junctions were not periodic, confirming what we had suspected: Two junctions were too special, not a reliable indicator of what to expect in larger arrays.

As the computer continued to churn, a different pattern materialized. The dots were tracing out a curve, not a blob, and their motion was meticulous, confined to a razor-thin path, extending it, filling it out. All the panels were showing different versions of the same basic structure: a distorted, triangular loop with rounded corners. I wondered if maybe I'd chosen a pathological starting point by accident, so I tried many other initial conditions. My jaw dropped when I saw the results. Every starting point gave rise to its own rounded triangle, and all the separate triangles fit neatly inside one another like Russian dolls.

This structure was incredible. It meant that the equations contained a secret symmetry, a hidden regularity that must be causing this order. I'd never seen anything like it. Every trajectory had an unimaginably vast, 10-dimensional landscape to explore, with the potential to wander up and down, front and back, left and right, and in 7 other dimensions that we don't even have words to describe, and yet they all did nothing of the sort. It was as unlikely as walking a tightrope forever and never falling off. Something was confining the solutions to a slice of all possibilities. It didn't even matter when I added more junctions to the array: 20, 50, 100—all yielded the same Russian-doll pattern of nested triangles. When I told Kurt the news, he was every bit as flabbergasted. Either the computer was playing tricks on us, or there was something unprecedented about the mathematics of Josephson arrays.

Over the next four years, many of us became obsessed with the mystery. Kurt and his student Steve Nichols ran computer simulations on a wider class of arrays, and kept detecting the same telltale signs of astonishing order. Jim Swift, a mathematician at Northern Arizona University and a friend of Kurt's from graduate school, dreamed up an ingenious way to approximate the equations that governed the dynamics of these arrays, replacing them with so-called averaged equations that were much easier to analyze but that nevertheless retained the essence of the original equations. (Like all puzzle solvers, mathematicians often resort to approximations when a problem seems too tough to approach head-on, at least at first.) By simplifying the problem, Jim opened the door to its mathematical analysis. Following his lead, my student Shinya Watanabe found the Russian-doll structure lurking in the solutions to Jim's averaged equations; then, in an analytical tour de force, he went on to prove that much of the same structure was latent in the original, unaveraged circuit equations. The upshot was the discovery of a new "integrable system," a rare jewel in mathematics. It has no particular application, at least not that we know of. It's more like finding a pretty shell on the beach.

One of the most wonderful things about curiosity-driven research—aside from the pleasure it brings—is that it often has unexpected spin-offs. The techniques developed by Jim and Shinya allowed us, for the first time, to tackle the dynamics of Josephson arrays in the more realistic case where the junctions are not identical. Engineers had never been able to analyze disordered arrays, though they knew full well that real junctions always differ by a few percent in their electrical properties; there's no way to fabricate them more uniformly than that with present manufacturing technology. The variability of the junctions limits their usefulness in arrays, because it opposes the coherent operation that engineers seek. When such arrays are driven by external currents, they are found to be temperamental: At currents below some threshold, they remain incoherent, with all the junctions oscillating at random phases such that their voltages interfere destructively and cancel out; but when the threshold is crossed, the array spontaneously synchronizes. To try to make sense of this behavior, Kurt and I (in collaboration with his friend Pere Colet) used Jim's averaging technique to massage the equations into a more manageable form.

There, staring us in the face, was the Kuramoto model—an enigma like the monolith in *2001: A Space Odyssey,* buried under the soil, waiting for us apes to find it, beckoning, the key to sync. Until now, the Kuramoto model had been thought to be nothing more than a convenient abstraction, the simplest way to understand how groups of dissimilar oscillators could spontaneously synchronize, and under what circumstances. It was born out of pure imagination, concocted as a caricature of biological oscillators: crickets, fireflies, cardiac pacemaker cells. Now here it was, unearthed, in the dynamics of superconducting Josephson junctions. It reminded me of that wonderful feeling that Einstein talked about, the recognition of hidden unity.

Soon after we published these results, I received a letter postmarked Kyoto, Japan, handwritten in graceful script. "I was surprised and really delighted," wrote Yoshiki Kuramoto. "I didn't have a slightest idea that my simple model could ever find any example in real physical systems."

The Kuramoto model has always been a solution waiting for a problem. It was never intended as a literal description of anything, only as an idealized model for exploring the birth of spontaneous order in its simplest form. Yet its newfound connection to Josephson arrays immediately explained why these devices should synchronize abruptly. The phase transition was fundamentally the same as the one that Winfree had discovered in his model of biological oscillators, and that Kuramoto had later formalized so elegantly in his solvable model. Experts on Josephson junctions had seen this transition in their own computer simulations, years earlier, but without a theoretical basis for understanding it, it had never attracted attention (illustrating the adage that you should never trust a fact until it's been confirmed by theory).

Since 1996, the Kuramoto model has turned up in other physical settings, from arrays of coupled lasers to the hypothesized oscillations of the wispy subatomic particles called neutrinos. We may be catching the first glimpses of a deep unity in the nature of sync. Whether there will be any practical applications remains to be seen. Given how many diseases are related to synchrony and its disruption (epilepsy, cardiac arrhythmias, chronic insomnia) and how many devices rely on synchrony (Josephson and laser arrays, electrical power grids, the global positioning system), it seems safe to say that a deeper understanding of spontaneous sync is bound to find practical benefit.

The widespread occurrence of the Kuramoto model raises the question of why this particular mathematical structure should be so common. To be honest, it probably isn't all that common. I have focused on it because it is the only case of spontaneous synchrony we understand well. On theoretical grounds, one can show that it arises only whenever four specific conditions are met, and is not expected otherwise. First, the system in question must be built from an enormous number of components, each of which is a self-sustained oscillator.

That is already a strong constraint. The individual elements must have extremely simple dynamics: pure rhythmicity along a standard cycle, without chaos or turbulence or anything complicated, just repetitive motion. Second, the oscillators must be weakly coupled, in the sense that the state of each oscillator can be characterized by its phase alone. If the coupling is strong enough to distort any oscillator's amplitude significantly, the Kuramoto model will not apply. The third condition is the most restrictive: Each oscillator must be coupled equally strongly to all the others. Very few systems in nature are literally like that. Oscillators normally interact most strongly with their neighbors in space, or with a collection of virtual neighbors defined by a network of mutual influence. Finally, the oscillators must be nearly identical, and the amount of dispersion in their properties should be comparable to the weakness of their coupling.

Given all these conditions, the dynamics of the Kuramoto model and its relatives might start to seem self-evident. Yet the sudden onset of synchrony still comes as a surprise. Even after sync breaks out, we often lack intuition about it, especially about how it could have occurred so abruptly on its own— as illustrated recently by the Millennium Bridge fiasco.

The Millennium Bridge was supposed to be the pride of London. Erected at a cost of over $27 million, the elegant, avant-garde footbridge was London's first new river crossing in more than a century, linking the City and St. Paul's Cathedral on the north bank of the Thames to the Tate Modern museum on the south. Its design was radical—the world's flattest suspension bridge, a sinuous ribbon 320 meters long, with low-slung outriggers and slender steel cables stretched taut across the river. The concept grew out of an unusual collaboration between the engineering firm Ove Arup, the architect Lord Norman Foster, and the sculptor Sir Anthony Caro. "A blade of light," Lord Foster dubbed it, imagining its appearance when illuminated at night. "I remembered going to the pictures and seeing Flash Gordon. As he got to the edge of an abyss he hit a button and this light-bridge appeared. That's what we wanted to create, something as close to flying as possible." Though the engineers at Arup were responsible for building the structure and ensuring its soundness, Lord Foster and Sir Anthony seemed happy to share the credit at the televised inaugural led by the Queen.

The bridge opened to the public on a sunny Saturday, June 10, 2000. As soon as police gave the word, hundreds of excited Londoners surged onto the deck from both ends. Within minutes it began to wobble, 690 tons of steel and aluminum swaying in a lateral S-shaped vibration like a snake slithering on the ground. Alarmed pedestrians clung to the handrails to steady themselves but the wobbling grew ever more violent, ultimately reaching deflections of 20 centimeters from side to side.

Roger Ridsdill-Smith, one of the young engineers at Arup who came up with the innovative design, looked over at the police crowd-controllers. This wasn't supposed to be happening. His mind raced—nothing like this had been predicted by the computer simulations, the safety assessments, the wind-tunnel experiments. The bridge was safe, he was sure of that. It couldn't possibly collapse like the Tacoma Narrows Bridge, the infamous "Galloping Gertie" preserved on grainy old film strips, caught in its death throes, twisting in the wind, crumbling in a fit of torsional oscillations. Still, something was causing the bridge to resonate. Police restricted access to the bridge, but the swaying continued. Panicked and humiliated, the authorities closed the Millennium Bridge on Monday, June 12, just two days after it opened.

Critics of the original design snorted about the blade of light's comeuppance. Lord Foster was no longer so eager to take credit; besieged by reporters, he extruded some ill-tempered words about his engineering collaborators. Arup, the engineering firm, immediately set about testing the bridge's vibrational characteristics to determine what had gone wrong. They attached huge shaking machines to the bridge and systematically wiggled it at a controlled range of frequencies. When the bridge was shaken horizontally at about 1 cycle per second, it slithered back into the S-shaped wobble seen on opening day.

That was an important clue. One cycle per second is half the frequency of normal human walking. All bridge designers know that people walk at a pace of about two strides per second, but the main effect of these repetitive footfalls is to create a vertical force, not a sideways one, so that couldn't be the cause of the lateral wobbling. Suddenly the engineers knew the culprit. People do create a small sideways force with each step—you push off one way when you plant your right foot, and the other way when you plant your left one. That alternating sideways force oscillates at half the stride frequency, one cycle per second, not two. No one had ever thought to worry about that; it wasn't part of the standard code for bridge designers in the United Kingdom. In any case, the sideways force is small, and since there's normally no coordination between people in large groups, all the leftward and rightward forces occur at random times and therefore tend to cancel each other out. But if for some reason everyone were stepping in sync, all the sideways forces would add up and become concentrated. That could definitely cause trouble.

The engineers went back and looked at the television news footage of opening day and saw that was exactly what had happened. As the bridge swayed, the pedestrians unconsciously adjusted their pace to walk in time with the lateral movement. This exacerbated the vibration, which impelled more people to lose their balance and simultaneously swing to the same side, reinforcing their synchrony and aggravating the vibration still further. It was this chain reaction—the positive feedback between the people and the bridge—that no one had ever anticipated, and that triggered the wobbling of the Millennium Bridge.

This resonance effect is different from the famous one that requires soldiers to break step before they cross a bridge, to avoid exciting dangerous vibrations in it. Soldiers arrive at the bridge in sync, whereas the pedestrians were strolling at random; the designers had no reason to expect them to spontaneously coordinate their footfalls. They'd prudently considered the possibility that a pack of vandals might deliberately jump up and down in sync, and designed the bridge to withstand that insult, but it had never dawned on them that a crowd of 2,000 civic-minded people could inadvertently synchronize their strolling.

It's still unclear what initiated the synchrony on opening day. The best guess is that a nucleus of sync was created by accident: Once the crowd is large enough, there's a chance that at some stage, enough people will step in sync by accident that a critical threshold will be crossed and the bridge will begin to wobble slightly. Once that happens, the feedback effect kicks in and reinforces the swaying.

Arup's later investigations showed that this kind of chain reaction is possible only if the bridge is very long, flexible, and crowded—the volatile mix of ingredients that combined on the Millennium Bridge that day. In particular, they found there's no sign of trouble if there are fewer people than the threshold number. It's not as if the bridge shakes a little for a small number of people and gradually builds up as the numbers increase. Either it doesn't shake at all, or it wobbles violently and without warning, once the threshold is crossed. Like the straw that breaks the camel's back, the onset of wobbling is a nonlinear phenomenon.

In fact, it sounds very much like the phase transition predicted by the models of Winfree and Kuramoto. Just as the theories suggest, the oscillators (in this case, people's footfalls) are incoherent below threshold. The forces they exert cancel each other out. They remain incoherent even as the coupling between them is increased; the coherence does not grow gradually. Then suddenly, once the coupling exceeds a certain threshold (because there are enough people on the bridge to shake it sufficiently), synchrony breaks out cooperatively.

We can see another conceptual unity here. The Millennium Bridge was a case of sync induced by weak coupling through an intermediate. That theme has been an undercurrent throughout the past few chapters. The pedestrians' interactions were mediated through the vibrations they induced in the bridge, in much the same way that Huygens's pendulums felt each other by shaking the board from which they were both suspended. In superconductivity, Cooper pairs form because electrons deform the atomic lattice slightly; that deformation provides a weak attraction between them, just as a bowling ball rolling on a waterbed tends to pull another along in its wake. Even in a series array of

Josephson junctions, the same mechanism is present: The junctions interact only because of the electrical oscillations they induce in the load. The individual oscillators in all four cases are completely different—electrons, pendulums, high-tech devices, people—but the synchronization mechanism is essentially the same.

The crux of this explanation was confirmed by Arup's engineers after several months of careful testing, involving not only their huge mechanical shakers but also controlled experiments with people walking across other bridges and laboratory studies of individual people balancing themselves on wobbly footing. But incredibly, just two days after the bridge closed, and before any studies had been conducted, a reader of London's *Guardian* had already arrived at the correct explanation. On June 14, 2000, the following letter to the editor appeared:

Out of step on the bridge

Wednesday June 14, 2000
The Guardian

The Millennium Bridge problem (Millennium bug strikes again, June 13) has little to do with crowds walking in step: It is connected with what people do as they try to maintain balance if the surface on which they are walking starts to move, and is similar to what can happen if a number of people stand up at the same time in a small boat. It is possible in both cases that the movements that people make as they try to maintain their balance lead to an increase in whatever swaying is already present, so that the swaying goes on getting worse.

Is it true that "the bridge is never going to fall down," or at any rate get damaged, as a result of the swaying? That has been said about bridges before, and those responsible for this one need to understand, before making such pronouncements, that the problem involves more than engineering principles.

The author, with his confident mix of scientific insight and contempt for received wisdom, signed his name:

Prof. Brian Josephson
Department of Physics
University of Cambridge

III
EXPLORING SYNC

Seven

SYNCHRONIZED CHAOS

H<small>E DIDN'T GIVE THE IMPRESSION OF BEING</small> a revolutionary. A small, modest man, seventyish, prone to speaking in a monotone, Ed Lorenz looked and acted more like a quiet country person, like a farmer you might see at a roadside stand in Maine. I'd often see him when I ate dinner at the MIT cafeteria in Walker Memorial. He'd hobble in with his wife, holding hands, holding canes with their free hands. Every time I taught my chaos course, we'd go through the same ritual each year, and I'd come to look forward to it. I'd call up Professor Lorenz and invite him to come give a guest lecture to the class. He'd say, with genuine puzzlement, as if it were an open question, "What should I talk about?" And I'd say, How about the Lorenz equations? "Oh, that little model?" And then, as predictable as the seasons, he'd show his face to my awestruck class, and tell us not about the Lorenz equations but about whatever he was working on then. It didn't matter. We were all there to catch a glimpse of the man who'd started the modern field of chaos theory.

"That little model" had changed the direction of science forever. In 1963, while trying to understand the unpredictability of the weather, Lorenz wrote down a set of three differential equations, nonlinear ones, but not horrible-looking. In fact, to a mathematician or physicist, they looked deceptively simple, like the standard exercises found in textbooks. I could solve that, you'd think to yourself. But you couldn't. No one could. The solutions to the Lorenz equations behaved like nothing mathematics had ever seen. His equations generated chaos: seemingly random, unpredictable behavior governed by nonrandom, deterministic laws.

At first, nobody noticed the new arrival. Lorenz's paper "Deterministic Nonperiodic Flow," buried on pages 130 to 141 of the *Journal of the Atmospheric Sciences,* was cited only about once a year for the first decade of its existence. But once the chaos revolution was in full swing, in the 1970s and 1980s, the little model averaged a hundred citations a year.

The first wave hit when a few scientists in diverse fields began to realize that they were all seeing manifestations of the same mysterious phenomenon. Ecologists stumbled upon chaos in a simple model for the dynamics of a wildlife population. Instead of leveling off or repeating in cycles, the simulated population unexpectedly boomed and crashed erratically from one generation to the next, even though there was nothing random in the model itself. Astronomers were perplexed by their measurements of the rotational motion of Hyperion, a small, potato-shaped moon of Saturn; instead of spinning on one axis like most satellites, it tumbled haphazardly, as if doing drunken somersaults. Physicists took time off from pondering quarks and black holes and began to pay attention to more mundane phenomena that they'd previously dismissed as annoyances: the fitful pulsations of unstable laser beams, the noisy voltage oscillations of certain electrical circuits, even the dripping of leaky faucets. All of these, it turned out, were to become icons of chaos. Ironically, a handful of pure mathematicians starting with Henri Poincaré had known about chaos for 70 years, but almost no one else could decipher their jargon or understand their abstractions, so their ideas had little impact outside their small priesthood.

And that's typical of the obstacles facing the development of any cross-disciplinary science. Most scientists work comfortably in their narrow specialties, walled off from their intellectual neighbors by barriers of language, taste, and scientific culture. Lorenz was not like that. He was a meteorologist whose first love had been mathematics. There were people like him in every field, mavericks within their own communities. What they had in common was a feeling for dynamics, for flow, for hidden patterns and symmetries, and above all, for the lure of the darkest corner of theoretical science: the realm of non-linear problems.

The mathematician Stanislaw Ulam once said that calling a problem non-linear was like going to the zoo and talking about all the interesting non-elephant animals you see there. His point was that most animals are not elephants, and most equations are not linear. Linear equations describe simple, idealized situations where causes are proportional to effects, and forces are proportional to responses. If you bend a steel girder by two millimeters instead of one, it will push back twice as hard. The word *linear* refers to this proportionality: If you graph the deflection of the girder versus the force applied, the relationship falls on a straight line. (Here, *linear* does not mean sequential, as in "linear thinking," plodding along, one thing after another. That's a different use of the same word.)

Linear equations are tractable because they are modular: They can be broken into pieces. Each piece can be analyzed separately and solved, and finally all the separate answers can be recombined—literally added back together—to give the right answer to the original problem. In a linear system, the whole is exactly equal to the sum of the parts.

But linearity is often an approximation to a more complicated reality. Most systems behave linearly only when they are close to equilibrium, and only when we don't push them too hard. A civil engineer can predict how a skyscraper will sway in the wind, as long as the wind is not too strong. Electrical circuits are completely predictable—until they get fried by a power surge. When a system goes nonlinear, driven out of its normal operating range, all bets are off. The old equations no longer apply.

Still, you shouldn't get the idea that nonlinearity is dangerous or even undesirable. In fact, life depends on nonlinearity. In any situation where the whole is not equal to the sum of the parts, where things are cooperating or competing, not just adding up their separate contributions, you can be sure that nonlinearity is present. Biology uses it everywhere. Our nervous system is built from nonlinear components. The laws of ecology (to the extent we know them) are nonlinear. Combination therapy for AIDS patients—drug cocktails—are effective precisely because the immune response and the viral population dynamics are both nonlinear; the three drugs in combination are much more potent than the sum of the three of them taken separately. And human psychology is absolutely nonlinear. If you listen to your two favorite songs at the same time, you won't get double the pleasure.

This synergistic character of nonlinear systems is precisely what makes them so difficult to analyze. They can't be taken apart. The whole system has to be examined all at once, as a coherent entity. As we've seen earlier, this necessity for global thinking is the greatest challenge in understanding how large systems of oscillators can spontaneously synchronize themselves. More generally, all problems about self-organization are fundamentally nonlinear. So the study of sync has always been entwined with the study of nonlinearity.

The synergistic character of nonlinear systems is also what makes them so rich. Every major unsolved problem in science, from consciousness to cancer to the collective craziness of the economy, is nonlinear. For the next few centuries, science is going to be slogging away at nonlinear problems. Starting in the 1960s and 1970s, all of the pioneers of sync—people like Wiener, Winfree, Kuramoto, Peskin, and Josephson—were already blazing one path up the mountain, on the trail of spontaneous order in enormous systems of oscillators. With the rise of chaos theory, an army of new allies had joined the quest, clambering up a separate trail but headed for the same peak.

Nonlinear problems had always been opaque. It was for this reason that Lorenz's headway on the problem of chaos was so encouraging. Now, suddenly, it became clear that even the simplest nonlinear systems could display very complicated behavior, much more complicated than anyone had realized. That might sound like a pessimistic conclusion, but it raised the hope that some seemingly random phenomena might harbor a deeper lawfulness within.

And then came the second wave of chaos theory, which revealed that chaos itself, belying its misleading name, contained a stunning new kind of order. The pivotal discovery was made by the physicist Mitchell Feigenbaum, who showed that there are certain universal laws governing the transition from regular to chaotic behavior. Roughly speaking, completely different systems can go chaotic in the same way. His predictions were soon confirmed in experiments on electronic circuits, swirling fluids, chemical reactions, semiconductors, and heart cells. It was as if the old Pythagorean dream had come true: The world was not made of earth, air, fire, and water—it was made of number. Feigenbaum's laws transcended the superficial differences between heart cells and silicon semiconductors. Different materials, the same laws of chaos. Other universal laws would soon be discovered. The logjam seemed to be broken.

It was a euphoric time for nonlinear science. *Chaos*—the word itself was cool. The field was touted by some as the third great revolution of twentieth-century physics, along with relativity and quantum mechanics. It had penetrated some of the mysteries of nonlinearity for the first time, and established links between fields that previously seemed unrelated. In 1987, James Gleick's best-selling book *Chaos* brought chaos theory to the masses, with stories of heroes like Lorenz and Feigenbaum, an intense, chain-smoking genius with Beethoven hair, wandering the streets of Los Alamos in the middle of the night, looking for the secret of turbulence. And then, when Jeff Goldblum played a chaos theorist in *Jurassic Park,* dressed in leather and looking like a rock star, chaos had truly arrived—especially after he demonstrated the butterfly effect on Laura Dern's hand.

The butterfly effect came to be the most familiar icon of the new science, and appropriately so, for it is the signature of chaos. The phrase comes from the title of a 1979 paper by Lorenz called "Predictability: Does the Flap of a Butterfly's Wings in Brazil Set Off a Tornado in Texas?" The idea is that in a chaotic system, small disturbances grow exponentially fast, rendering long-term prediction impossible.

A depressing corollary of the butterfly effect (or so it was widely believed) was that two chaotic systems could never synchronize with each other. Even if you took great pains to start them the same way, there would always be some infinitesimal difference in their initial states. Normally that small discrepancy would remain small for a long time, but in a chaotic system, the error cascades and feeds on itself so swiftly that the systems diverge almost immediately, destroying the synchronization. Unfortunately, it seemed, two of the most vibrant branches of nonlinear science—chaos and sync—could never be married. They were fundamentally incompatible.

Plausible as it sounds, the argument outlawing synchronized chaos is now known to be wrong. Chaos *can* sync.

This startling phenomenon was discovered in the early 1990s, and with it came another change of perspective about chaos itself. Traditionally, chaos had been viewed as a nuisance, something to be suppressed and engineered away. Later, in the heyday of the revolution, chaos became a celebrated curiosity. Its pervasiveness in the natural world was recognized, and its hidden order exposed. No one knew whether it was good for anything, but that didn't matter. It was fascinating for its own sake. Now, with the discovery of synchronized chaos, the sea changed again. Overnight, chaos promised to be useful. Physicists and engineers dreamed of ways to harness its remarkable properties to do potentially practical things, like scramble cell-phone calls and other wireless forms of communication to prevent eavesdroppers from intercepting them.

The discovery of synchronized chaos also enriched our understanding of sync itself. In the past, sync had always been associated with rhythmicity. The two concepts are so tightly linked that it's easy to overlook the distinction between them. Rhythmicity means that something repeats its behavior at regular time intervals; sync means that two things happen simultaneously. The confusion occurs because many synchronous phenomena are rhythmic as well. Synchronous fireflies not only flash in unison, they also flash periodically, at fixed intervals. Cardiac pacemaker cells fire in step, and at a constant rate. The moon turns once as it orbits Earth; both its spin and its orbit follow cycles that repeat themselves regularly.

But we all know that, at least in principle, sync can be persistent without being periodic. Think of the musicians in an orchestra. All the violins come in at the same time, and stay in sync throughout. Yet they are not periodic: They do not play the same passage over and over again. Or imagine the world champions in pairs figure skating. Their graceful movements occur in tandem, but are ever inventive, never repetitive.

These displays of sync without cycles impress us, delight us, sometimes even move us. They seem to require intelligence and artistry, which is why the discovery of synchronized chaos was so astonishing: It demonstrated that mindless things can pull off a primitive version of the same feat. Purely mechanical systems can glide along unpredictably while remaining in perfect concert.

To understand how synchronized chaos works, the first step is to understand chaos itself. Unfortunately, many of us begin with faulty preconceptions about what chaos is like. (Incidentally, the same is *not* true of periodicity. We instinctively understand it correctly. All the cycles around us—the beating of a heart, the ticking of a grandfather clock, the changing of the seasons, the insufferable *beep-beep-beep* of a truck backing up—give an accurate sense of what periodicity really means. You can feel the rhythmic pounding of a drum in the pit of your stomach as the parade marches by. Now we need to develop the same kind of visceral feel for chaos.)

Part of the confusion stems from the word itself. In colloquial usage, *chaos* means a state of total disorder. In its technical sense, however, *chaos* refers to a state that only appears random, but is actually generated by nonrandom laws. As such, it occupies an unfamiliar middle ground between order and disorder. It looks erratic superficially, yet it contains cryptic patterns and is governed by rigid rules. It's predictable in the short run but unpredictable in the long run. And it never repeats itself: Its behavior is nonperiodic.

The chaos governed by the Lorenz equations, for example, is vividly illustrated by a strange and beautiful contraption, a desktop waterwheel designed by Willem Malkus, one of Lorenz's former colleagues at MIT. It's intended as a pedagogical aid to give students an image of chaos in action. The original low-tech device, designed by Malkus and his colleague Lou Howard, was a lazy

Susan with a dozen leaky paper cups attached to its rim in the manner of chairs on a Ferris wheel. As Malkus told me, this prototype was a "messy affair"—as water was poured in from a watering can to set the wheel in motion, it would slowly drain through the cups and spill all over the table and floor.

His improved waterwheel, on the other hand, is a completely self-contained machine.

A plastic wheel, about a foot in diameter, rotates in a plane tilted slightly from the horizontal (unlike an ordinary waterwheel, which rotates in a vertical plane). With the flip of a switch, water is automatically pumped up into an overhanging manifold (a perforated hose) and then sprayed out through dozens of small nozzles into separate chambers around the rim of the wheel (the counterpart of the cups in the low-tech version). At the bottom of each chamber, the water leaks out through a pinhole and collects in a common reservoir underneath the wheel, where it is pumped back up through the nozzles. This recirculation scheme provides a steady inflow of water.

When you turn on the machine, nothing much happens at first. The wheel is motionless. The water makes a pleasant gurgling sound as it fills the chambers; meanwhile, those chambers are draining, but at a slower rate. Once the chambers get too full, the wheel becomes top-heavy and starts to swing around in one direction, like an inverted pendulum falling over. That rotation carries a new set of chambers under the manifold while simultaneously transporting some of the filled ones out from under the nozzles. Soon you start to feel like you're seeing the pattern: The wheel is consistently rotating in one direction, say counterclockwise. After another minute, however, the rotations become increasingly sluggish, barely making it over the top, as the wheel becomes more and more imbalanced from the lopsided placement of water around its rim. As the wheel strains to make one last revolution, it doesn't quite succeed and slows to a halt, then reverses direction, now turning clockwise. Wait a little longer and soon the wheel settles into its remarkable steady-state behavior: a haphazard sequence of clockwise and counterclockwise turns, punctuated by reversals at unpredictable times. It might spin three times clockwise, then once counterclockwise, four times one way, seven times the other. There's no discernible

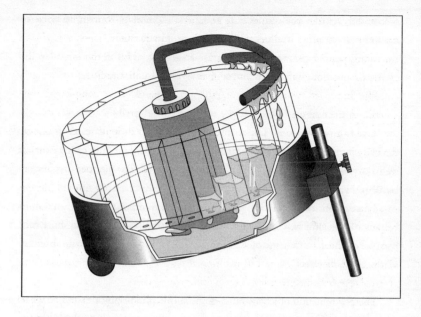

trend. The long-term behavior is nonperiodic. The motion never damps out and never repeats.

What's so surprising here is that the waterwheel turns erratically, even though there's nothing erratic about the way it is being driven. The water is pumped in at a steady rate. Yet the wheel can't seem to make up its mind. What's even more disconcerting is that the behavior is not reproducible. The next time you turn on the waterwheel, its pattern of reversals will be different. If you take tremendous care to ensure that everything is almost the same as it was last time, its motion will track for a while but then diverge, yielding a completely unrelated sequence of turns and reversals thereafter.

Of course, if you started the wheel *absolutely* the same way, it would repeat. That's what it means to be deterministic: The current state determines the future state uniquely. The motion of the wheel is governed by deterministic equations—Newton's laws of motion and the laws of fluid mechanics—so in principle, if you knew all the variables initially, you could predict the wheel's motion forever into the future. The equations themselves contain no noise or randomness or other sources of uncertainty. Furthermore, if you solve the equations on a computer, using the same starting values for all the variables, the predicted outcome will be the same every time. In that sense, everything is reproducible.

But in the real world outside the computer, the variables are never exactly the same from run to run. The slightest difference—a drop of water in one of the chambers, left over from the previous experiment, or a puff of air exhaled by an overexcited spectator—will alter the motion of the wheel, at first imperceptibly, but very soon with incalculable consequences.

These, then, are the defining features of chaos: erratic, seemingly random behavior in an otherwise deterministic system; predictability in the short run, because of the deterministic laws; and unpredictability in the long run, because of the butterfly effect.

The phenomenon of chaos raises some subtle philosophical issues that can trick the unwary. For example, a few of my students have pooh-poohed the butterfly effect as obvious. We all know that little things can make a big difference in the course of our own lives, and even in the lives of nations. With so many complexities, so many variables unaccounted for, insignificant events can sometimes trigger disproportionate chain reactions. Think of the ancient verse about the downfall of a kingdom:

> For want of a nail, the shoe was lost;
> For want of a shoe, the horse was lost;
> For want of a horse, the rider was lost;
> For want of a rider, the battle was lost;
> For want of a battle, the kingdom was lost!

But what was not widely appreciated until chaos theory was that similar cascades can afflict even the simplest systems: waterwheels and tumbling moons and dripping faucets, mechanical systems where all the laws are known and there are only a few variables. Even there, the seeds of chaos lie dormant, ready to unfold their surprises.

Another subtlety: In chaos, every point is a point of instability. It's worse than the quandary faced by Robert Frost's traveler in "The Road Not Taken"— a life ruled by chaos is even more precarious. Every moment would be a moment of truth. Every decision would have long-term consequences that would alter your life beyond recognition. Button your shirt starting from the top instead of the bottom, and there's no telling how differently things might turn out, years later. (Our lives might actually be like that; we get to follow only one trajectory, so we have no way of knowing what fate would hold for the others, where we start buttoning from the bottom first. But to retain a measure of sanity, one has to believe that nearly all such decisions are inconsequential. This dilemma was explored in the film *Sliding Doors,* which depicts two radically different versions of a woman's life, depending on whether or not she catches a subway train before the doors slide closed.)

In contrast to chaotic systems, rhythmic systems don't show such inordinate sensitivity to small disturbances. Smack a metronome—it stutters, but then resumes its relentless *ticktock*. The timing is off from where it would have been, but that deviation does not grow as time passes. We can see this more clearly if we imagine two identical metronomes, initially in step. Disturb one of them; after it recovers, it will be out of step from the other by a fixed interval. The discrepancy does not grow. More generally, when a nonchaotic system is disturbed slightly, the disturbance either doesn't grow at all or else grows very mildly, increasing in proportion to how much time has passed. One says that the errors grow no faster than linearly in time.

The important point here is a quantitative one. The linear growth of errors implies that nonchaotic systems are predictable, at least in principle. The tides, the return of Halley's comet, the timing of eclipses: All of these are strongly rhythmic and hence predictable, because tiny disturbances do not mushroom into major forecasting errors. To forecast a nonchaotic system twice as long, you just measure its initial state twice as precisely. To go three times longer, you improve your measurements threefold. In other words, the horizon of predictability also increases linearly: that is, in direct proportion to the precision with which the initial state is known.

Chaotic systems, on the other hand, behave in a radically different way, and it is here that we begin to grasp the truly demoralizing implications of the butterfly effect. The amount of time we can successfully predict the state of a chaotic system depends on three things: how much error we're willing to tolerate in the forecast; how precisely we can measure the initial state of the system; and a time scale that's beyond our control, called the Lyapunov time, which depends on the inherent dynamics of the system itself.

Roughly speaking, we can only predict for an amount of time comparable to the Lyapunov time; after that, the errors in the measurement of the true initial state have snowballed so much that they exceed the allowable tolerance. By lowering our standards or improving our initial measurements, we can always predict longer. But the rub is the obstinate way the predictability horizon depends on the initial precision: If you want to predict twice as long yet still achieve the same accuracy, it will now cost you not twice the effort but ten times as much. And if you are ambitious and want to predict three times longer, that will cost a hundred times the effort; four times longer, a thousand times the effort, and so on. In a chaotic system, the required precision in the initial measurement grows *exponentially*, not linearly.

That's devastating. It means, in practice, that you can never predict much longer than a small multiple of the Lyapunov time, no matter how good your instruments become. The Lyapunov time sets a horizon beyond which acceptable prediction becomes impossible. For a chaotic electrical circuit, the horizon is something like a thousandth of a second; for the weather, it's unknown but seems to be a few days; and for the solar system itself, five million years.

It's because the horizon is so long for the solar system that the motions of the planets seem utterly predictable to us today; and on the time scales of a human life, or even of the whole history of astronomy, they *are* predictable. When we calculate planetary alignments hundreds of years into the past or the future, our predictions are reliable. But any claims about the positions of the planets 4 billion years ago, at the dawn of life on Earth, would be meaningless.

. . .

The final subtlety about chaos has to do with a strange kind of order lurking within it. Chaos is not formless (again, despite the word's ordinary meaning). A hint of its underlying structure appears in the motion of the toy waterwheel, with its never-ending succession of rotations and reversals; although the sequence never repeats in detail, its overall character stays the same. Chaos has an essence, a quality that never changes.

When Lorenz was analyzing his little model back in the early 1960s, he happened upon the essence of chaos incarnate. It took the form of a shape, an alien thing, not quite a surface, but not a solid volume either. It wasn't easy to visualize it back then, long before the advent of modern computer graphics. Even after Lorenz saw it in his mind's eye, he struggled to find the words to convey its peculiar geometry. He described it as an "infinite complex of surfaces." Today we call it a "strange attractor."

Just as a circle is the shape of periodicity, a strange attractor is the shape of chaos. It lives in an abstract mathematical space called state space, whose axes represent all the different variables in a physical system. Lorenz's equations involved three variables, so his state space was three-dimensional. For the waterwheel—an exact mechanical analog of the Lorenz equations—one of the variables tells how fast the wheel is rotating and in which direction, while the other two characterize two particular aspects of how the water is distributed around the rim of the wheel. The values of these variables at any instant define a single point in state space, corresponding to a snapshot of the system at any one moment.

In the next moment, the state will change as the wheel rotates and the water flows in and redistributes itself. Moving from state to state, the system evolves, carried along by its own dynamics. Like the diagrams in an Arthur Murray dance lesson, the Lorenz equations are rules about where to step next. They define infinitesimal arrows at every point in state space. Wherever the state happens to be, it must follow the arrow at that point, which brings it immediately to a new point. Following the arrow there for an instant, it proceeds to the next point, and so on. As time passes and the values of the variables change, the point cruises through state space, tracing a continuous path called a trajectory, sailing like a comet through an imaginary realm that exists only in a mathematician's mind. The beauty of this idea is that it converts dynamics into geometry. A chaotic motion becomes a picture, something we can see, a static image we can stare at, inspect, and study.

What does chaos look like? The trajectory wanders around forever in state space. It can never close or cross itself, because chaos never repeats. Lorenz was able to prove that his trajectory was forever confined inside a certain large sphere, so it could never escape to infinity. Trapped in this ball, sentenced to wander around for eternity without ever intersecting itself, the trajectory must follow an extremely elaborate path. The temptation is to picture it like a tangled ball of string, a wild mess, with no structure to it.

But Lorenz's primitive computer graphics indicated that the trajectory was moving in a highly organized way, exploring only a tiny portion of the available space. In fact, it seemed to be attracted onto a particular surface—a delicate, microscopically thin membrane whose shape, ironically, resembled a pair of butterfly wings. The trajectory would loop around one of the wings, spiraling out from its center. Then, when it got close to the edge of the wing, it would dart over to the other wing, and begin spiraling out again. The trajectory made an unpredictable number of loops around each wing before jumping to the other one, just as the waterwheel makes an unpredictable number of rotations in one direction before reversing.

As Lorenz struggled to make sense of what the computer was telling him, he realized something had to be wrong. He knew the trajectory couldn't be confined to a surface: There would be no way for it to avoid crossing itself. The butterfly wings might look like a single surface, but they would actually have to be built from an infinite number of layers, packed so closely together that they would appear indistinguishable, like sheets of mica.

This infinite complex of surfaces—this strange attractor—embodies a new kind of order. Though the trajectory's motion is unpredictable in detail, it always stays on the attractor, always moves through the same subset of states. That narrowness of repertoire accounts for the order hidden in chaos and explains why its essence never changes.

To make these abstractions more concrete, visualize a strange attractor as a futuristic parking garage from the *Twilight Zone*. The garage is completely automated. While you sit passively behind the steering wheel, a towing apparatus hitches to your car and pulls you along through the garage. Like the Lorenz attractor, the garage has two wings; in this case, let's call them the east and west towers, both with infinitely many levels. When you're ready to go home, you flick the switch to turn on the towing apparatus. You descend for a while, and feel like you're making progress, though you're becoming dizzy from circulating around so many levels, when you suddenly sense that you have made no progress at all, and you have somehow arrived near the top level of the opposite tower. As the hell ride continues, you circulate endlessly, every so often making unpredictable switches between the two towers. You are destined to drive forever. Although you never get out, you never retrace your path. Occasionally you might return to the same level of the same tower, but never in quite the same place.

This is the fate of a trajectory on the Lorenz attractor. The towing apparatus is the differential equation; it is what determines the trajectory, both its speed and direction at every instant. The rules are completely deterministic: The trajectory's fate is determined by its initial conditions. By analogy, if you start from the same parking space in the garage, you and your car will be towed along the same path every time, speeding up and slowing down in the same way. The butterfly effect expresses itself through sensitive dependence on initial conditions: In the metaphor, if you and the person in the car next to you ask to leave at the same time, the towing apparatus takes you both on the same ride for a while—as you look out your windows at each other in desperation—but very soon you diverge, veering apart onto different levels and different fates. After that, your patterns of circulation around the two towers are completely uncorrelated. Nevertheless, the existence of the strange attractor ensures a certain kind of order. You're always stuck in the garage, circulating endlessly through the same kinds of states, though never in quite the same sequence.

Although the shape of chaos is nightmarish, its voice is oddly soothing. When played through a loudspeaker, chaos sounds like white noise, like the soft static that helps insomniacs fall asleep. In the autumn of 1988, when Lou Pecora began to daydream about using chaos to do something practical, he sensed a promise in its sound, where everyone before him had heard only a bland, meaningless hiss.

Pecora is a lighthearted, playful physicist with a self-effacing manner and an easy laugh. In the mid-1980s, he was working at the U.S. Naval Research Laboratory in Washington, studying positron annihilation in solids, spin waves in magnets, and other problems in solid-state physics. Looking for a change of direction, and intrigued by the excitement around chaos theory (the hottest topic in physics at the time), he tried to justify switching his research to such an esoteric subject. He knew his superiors would be more receptive if he could propose a way to harness chaos for practical benefit, military or otherwise. This pragmatic line of thought, natural as it seems in retrospect, had never occurred to anyone. Until then, chaos theory had been dominated by pure researchers, scientists fascinated by nature, not by engineering. Practical applications never crossed their minds.

Once Pecora asked himself whether chaos could be useful, he immediately thought of communications. Maybe secret messages could be shrouded in chaos, making them harder for an enemy to intercept and decode. An eavesdropper might not realize a message was being sent, and even if he did, he might have trouble pulling it out of the noise. To have any hope of making this encryption strategy work, Pecora knew he would first have to figure out how to synchronize a chaotic transmitter and receiver. All forms of wireless communication rely on synchronization. In the case of radio, for example, the process of tuning to a particular station locks the receiver to the frequency of the broadcast transmission. Once the sync is established, the song on the radio is extracted through a process called demodulation, which teases the music apart from the radio wave that carries it. The challenge now was to generalize the same idea to chaos, where the carrier would be a chaotic wave instead of a periodic one.

Pecora and his postdoctoral fellow Tom Carroll had no background in communications and both were still newcomers to chaos, so they weren't sure where to begin. The quickest way to gain insight seemed to be through computer simulations; at least they wouldn't have to invest weeks of effort building gadgets that didn't work anyway. So they tinkered on the computer, simulating various pairs of chaotic systems, linking them in different ways and hoping their wild fluctuations would fall in step. Nothing worked. The butterfly effect was too powerful. The simulated transmitter and receiver would stay together for a while, but soon came the inexorable drift and breakdown of synchrony.

Feeling discouraged, Pecora headed to Houston for the annual chaos meeting, a conference called Dynamics Days. He sat in the audience listening to the leaders in the field, trying to concentrate. But his mind kept wandering back to the synchronization problem. By the end of the conference, he was no closer than before. He caught a late plane home and arrived at his doorstep well past midnight, feeling exhausted and cranky. His wife and kids were fast asleep. Soon after he dozed off, he was awakened by the cries of his seven-month-old daughter, Anna, who needed a bottle. His wife volunteered to take care of her, but Pecora said no, he'd like to do it.

There in the stillness of his house, sitting peacefully with his baby daughter, cradling her in his arms, Pecora felt himself relax. His brain stopped buzzing. Later, when he returned to bed, the solution hit him. "I need to drive chaos with chaos—I need to drive the receiver with a signal that comes from the same kind of system." Although he worried that he'd forget the idea, he was too tired to climb out of bed to write it down.

When he woke up the next morning, the idea was still there. He couldn't wait to test it. He thought of trying it on the Lorenz equations, but he wasn't comfortable yet with solving differential equations on the computer, so he worked instead with a chaotic system that was easier to program. Pecora started the transmitter and receiver in different states, and then asked the computer to predict their behavior far into the future. As the numbers poured out, they bobbled erratically—the aperiodicity expected of chaos—but amazingly, their values converged toward each other. They were synchronizing. By driving the receiver with a chaotic signal transmitted from a duplicate of itself, Pecora had coaxed them to fluctuate in lockstep.

In technical terms, his scheme can be described as follows: Take two copies of a chaotic system. Treat one as the driver; in applications to communications, it will function as the transmitter. The other system receives signals from the driver, but does not send any back. The communication is one-way. (Think of a military command center sending encrypted orders to its soldiers in the field or to sailors at sea.) To synchronize the systems, send the ever-changing numerical value of one of the driver variables to the receiver, and use it to replace the corresponding receiver variable, moment by moment. Under certain circumstances, Pecora found that all the *other* receiver variables—the ones *not* replaced—would automatically snap into sync with their counterparts in the driver. Having done so, *all* the variables are now matched. The two systems are completely synchronized.

This description, although correct mathematically, does not begin to convey the marvel of synchronized chaos. To appreciate how strange this phenomenon is, picture the variables of a chaotic system as modern dancers. By analogy with the Lorenz equations, their names are x, y, and z. Every night they perform onstage, playing off one another, each responding to the slightest cues of the other two. Though their turns and gestures seem choreographed, they are not. On the other hand, they are certainly not improvising, at least not in the usual sense of the word. There's nothing random in how they dance, no element of chance or whimsy. Given where the others are at any moment, the third reacts according to strict rules. The genius is in the artfulness of the rules themselves. They ensure that the resulting performance is always elegant but never monotonous, with motifs that remind but never repeat. The performance is different from minute to minute (because of aperiodicity) and from night to night (because of the butterfly effect), yet it is always essentially the same, because it always follows the same strange attractor.

So far, this is a metaphor for a single Lorenz system, playing the role of the receiver in Pecora's communication scheme. Now suppose that time stands still for a moment. The laws of the universe are suspended. In that terrifying instant, x vanishes without a trace. In its place stands a new variable, called x'. It looks like x but it is programmed to be oblivious to the local y and z. Instead, its behavior is determined remotely by its interplay with y' and z', variables in a transmitter far away in another Lorenz system, all part of an unseen driver.

It's almost like the classic horror movie *Invasion of the Body Snatchers*. From the point of view of the receiver system, this new x would seem inscrutable. "We're trying to dance with x but suddenly it's ignoring all of our signals," think y and z. "I've never seen x behave like that before," says one of them. "Hey, x," the other whispers, "is it really you?" But x wears a glazed expression on its face. Just as in the movie, x has been taken over by a pod. It's no longer dancing with the y and z in front of it—its partners are y' and z', unseen doppelgängers of y and z, remote ones in the parallel universe of the driver. In that faraway setting, everything about x' looks normal. But when teleported to the receiver, it seems oddly unresponsive. And that's because the receiver's x has been hijacked, impersonated by this strange x' coming from out of nowhere. Sensitive souls that they are, y and z make adjustments and modify their footwork. Soon all becomes right again. The x, y, z trio glides in an utterly natural way, flowing through state space on the Lorenz attractor, the picture of chaotic grace.

But what is so sinister here, and so eerie, is that y and z have now been turned into pods themselves. Unwittingly, they are now dancing in perfect sync with their own doppelgängers, y' and z', variables they have never encountered. Somehow, through the sole influence of the teleported x', subtle information has been conveyed about the remote y' and z' as well, enough to lock the receiver to the driver. Now all three variables x, y, and z have been commandeered. The unseen driver is calling the tune.

Pecora's simulations showed that his scheme would work for equations in the computer. Now the question was whether it would work in the lab, where no two systems are ever identical or free from outside disturbances. He considered what chaotic system would be the most manageable experimentally. Electronic circuits seemed like the natural choice; they're fast, cheap, and easy to measure, giving plenty of data in a short time. Carroll agreed, and set to work trying to implement the Lorenz equations in electronic hardware. Almost immediately he found himself stymied. Those particular equations involve multiplication of x and y in one term, and x and z in another. To perform those operations electronically requires multiplier chips, and Carroll was finding that his off-the-shelf components were too unreliable to provide the accuracy he needed. A more serious problem was that the variables in the Lorenz equations change by a factor of 100,000 as the system evolves. That enormous dynamic range exceeds the capabilities of the power supplies typically used to drive electronic circuits. Reluctantly, Pecora and Carroll scrapped the idea of a Lorenz circuit.

In search of alternatives, they consulted with Robert Newcomb, an electrical engineer at the University of Maryland who had designed his own brand of chaotic circuits. Newcomb had let his imagination run free. He hadn't felt compelled to make circuits that mimicked Lorenzian waterwheels or lasers or any other physical system; he was just curious about chaos and wanted to explore it electronically. Carroll followed one of Newcomb's recipes and confirmed that the resulting circuit produced wild fluctuations in voltage and current. Plotted on an oscilloscope, the variables traced out a strange attractor—not the same as Lorenz's butterfly wings, but similar. The circuit was running at thousands of cycles per second and giving fast, beautiful chaos.

Now the synchronization scheme could be tested. Carroll built a second copy of the circuit, and wired it to the first one according to Pecora's rules. The theory predicted that the two circuits should both oscillate spasmodically but in perfect lockstep. To test for synchrony, Carroll set the oscilloscope to plot the receiver voltage y versus its transmitter counterpart y'. If the two fluctuating variables were equal, they should line up on a 45-degree diagonal (because when y is graphed horizontally, and y' is graphed vertically, the horizontal displacement y must equal the vertical displacement y' if their values are always equal). And since y and y' are always changing from moment to moment, they should race back and forth along that diagonal line but never depart from it.

Carroll flipped the switch to start the circuits. Within two milliseconds, the voltages leapt onto the diagonal and stayed there. "My hair still stands up when I think about it," Pecora told me. "I don't think I'll ever have a moment like that again. It's like seeing one of your kids being born."

Last day of classes, MIT, December 1991. I'd just given the final lecture in my chaos course, and everyone had filed out except for one student. Beaming with pride, he handed me a piece of paper crammed with handwritten formulas and theorems, all enclosed in perfect rectangular boxes. To prepare for the upcoming final exam, he'd distilled the whole course to a single page. Looking at his minuscule, machinelike printing, I knew what I was dealing with. Sure enough, Kevin Cuomo turned out to be one of the best students in the class.

Cuomo was doing his Ph.D. research on synchronized chaos in electrical circuits and their possible uses in communications. At the time, I was vaguely aware of Pecora and Carroll's 1990 paper, but had not studied it carefully. Cuomo wanted to tell me all about it—the words came tumbling out in a torrent—but then he jumped to his own work, and encouraged me to come see a circuit he'd built—the first electronic implementation of the Lorenz equations—and he also wanted me to check a mathematical proof he'd discovered, a demonstration of a new synchronization scheme that would always work for the Lorenz equations, no matter how the receiver and transmitter were started. He took a breath and continued: Pecora and Carroll had not offered any such proof, and that worried him—the reasoning wasn't especially difficult, just a standard application of Lyapunov functions, like we'd done in class—so maybe he was missing something?

As it turned out, Cuomo had done everything right. His proof was sound, and his circuit did simulate the Lorenz equations (to this day, Pecora cheerfully admits that he has no idea how Cuomo got it to work). But none of this is what Cuomo is known for. Over the course of the next year, he and his adviser Al Oppenheim would be the first to demonstrate that chaotic encryption was possible: Synchronized chaos really could be used to enhance the privacy of communications.

Their method is based on masking, the same strategy used (unsuccessfully and unforgettably) by the secretive couple in Francis Ford Coppola's movie *The Conversation*. Fearing that they are under surveillance, a man and woman walk around a busy town square and whisper to each other, trusting that the loud din of street musicians will hide their conversation. In Cuomo and Oppenheim's version, the background noise is provided by the hiss of electrical chaos, generated by the variable x from a Lorenz circuit. Before any message is sent to the receiver, x is added on top of it, to mask it. For good coverage, x must be much louder than the message (just as the street music needs to be much louder than the whispered conversation) over its entire range of frequencies. Of course, if the receiver can't disentangle the message from the mask, nothing has been accomplished. This is where synchronization comes in. Cuomo's scheme ensures that the receiver, when driven by the hybrid signal (message plus mask), will synchronize to the mask, but not to the message. In effect the receiver regenerates a clean version of the mask. Subtracting it from the hybrid signal reveals the message. The method confers privacy because an eavesdropper has no easy way to perform the same decomposition; he wouldn't know what to subtract, what part of the combined signal is mask and what part is message.

A year after he took my course, Cuomo returned to give a live demonstration of his encryption scheme to my latest crop of chaos students. First he showed us his transmitter circuit: a small board loaded with resistors, capacitors, operational amplifiers, and analog multiplier chips. The voltages x, y, z at three different points in the circuit were proportional to Lorenz's variables of the same names. When Cuomo graphed x against y on an oscilloscope, the familiar butterfly wings of the strange attractor appeared as a glowing, ghostly image on the screen. Then, by hooking the transmitter up to a loudspeaker, Cuomo enabled us to hear the chaos. It crackled like static on the radio. Next he grabbed another circuit board, a receiver built to match the transmitter, and connected them with an alligator clip in a strategic place. Using the oscilloscope again, he demonstrated that both circuits were now running in sync, offering the usual 45-degree diagonal test as evidence.

Cuomo brought the house down when he used the circuits to mask a message, which he chose to be a recording of the hit song "Emotions," by Mariah Carey. (One student, apparently with different taste in music, asked, "Is that the signal or the noise?") After playing the original version of the song, Cuomo played the masked version. Listening to the hiss, one had absolutely no sense that there was a song buried underneath. Yet when this masked message was sent to the receiver, its output synchronized almost perfectly to the original chaos, and after instant electronic subtraction we heard Mariah Carey again. The song sounded fuzzy but was easily understandable.

When Cuomo and Oppenheim's paper was published in 1993, their dramatic results came as no surprise to Lou Pecora. He and Tom Carroll had been toiling along the same lines for three years already, but they weren't allowed to say anything or publish what they'd found.

As early as the fall of 1989, once their chaotic circuits were successfully synchronizing, Pecora and Carroll had begun considering the problem of chaotic encryption. Lacking even a rudimentary background in communications or coding theory, they came up with a clumsy method, one that required sending two signals. One signal was used to establish synchrony between the receiver and the transmitter. The second was a hybrid, a mask with a message added to it at very low power. It's essentially the same strategy that Cuomo and Oppenheim proposed a few years later, though less elegant in the sense that Cuomo's method uses only one signal (x plus message) for double duty—it both establishes sync and carries the message. But the general idea is the same.

The Space Warfare group at the Naval Research Laboratory became interested in Pecora and Carroll's work, because of the potential it offered for new ways of encoding and encrypting satellite communications. They had been funding Carroll for the preceding year, and now wanted a closer look at what the physicists were up to. A senior officer told Pecora to keep quiet about the work until the Space Warfare people had a chance to evaluate it; they were going to send an outside expert to assess the circuit. Pecora was given strict instructions about how to behave. He and Carroll were not allowed to ask the expert anything: not who he worked for, not even his name. What should we call him? Pecora asked. "Call him Bill," said his superior. In private, Pecora and Carroll referred to him as Dr. X.

Dr. X turned out to be a young man, serious and competent, carrying a computer loaded with software for simulating analog circuits. He seemed unfamiliar with chaos theory, but he clearly understood the communications ideas, and managed to get his own simulations of the circuit running very quickly. Pecora and Carroll were later informed that Dr. X had concluded that their circuit performed as described, though he had doubts about whether it could be made digital and secure.

Other visitors from the Space Warfare group soon followed. Pecora, in his naïveté, bet one of them a beer that he could hide a sine wave in the chaos, and challenged the visitor to extract it. The visitor ran the circuits for a minute, measured the voltage waveforms, then did a computation called a fast Fourier transform to measure the strengths of all the component frequencies being transmitted. The sine wave stood out nakedly as a spike in the spectrum. Pecora realized then that he had a lot to learn about encryption.

The Space Warfare scientists concluded that this new scheme was interesting but hardly something the navy should depend on. Pecora and Carroll were finally given permission to disclose their results, but because they wanted to apply for a patent, their lawyer advised them to extend their silence about what they were doing. So they still didn't publish anything.

Space Warfare also put them in touch with a contact at the National Security Agency, the ultrasecretive arm of the government concerned with the making and breaking of codes. Pecora visited the agency headquarters and presented his results to an audience of cryptographers who listened attentively, but wouldn't respond to any of his questions. "It was like talking in a black hole," said Pecora. "Information goes in and none comes out." After the meeting, Pecora realized he'd forgotten something and needed to get back in touch with his contact at NSA. Having lost the phone number, he looked in the phone book, and was surprised to find a listing for this most clandestine of organizations. He dialed the number and reached an information desk. The conversation was reminiscent of a Monty Python sketch:

> "May I have the phone number for Colonel Y?"
> "I cannot confirm or deny that anyone named Colonel Y works here."
> "OK, how about if I give you my number and you tell him to call me back?"
> "I cannot confirm or deny that he works here."
> "This is the information desk, isn't it?"
> "Yes. What information would you like?"

· · ·

The early work on synchronized chaos led to a jubilant sense of optimism about the prospects for chaotic encryption, especially among physicists with no background in cryptography. It was common in the early 1990s to see papers in physics journals with hopeful titles about "secure" communications. But the experts knew better. From the beginning, Al Oppenheim cautioned Cuomo and me about hyping the results. "You must never call this method secure," he warned. "Secure means secure—unbreakable. We don't know if it's secure. It may give some low level of privacy, but that's all. Masking schemes are usually pretty easy to break."

For people using cellular phones, even a minimal level of privacy would be welcome. Princess Diana needed it when reporters intercepted her conversations with her lover James Gilbey, later publicized as the embarrassing "Squidgy" tapes. Prince Charles was caught speaking even more intimately to Camilla Parker Bowles in 1989. When Newt Gingrich and his lawyers were discussing the ethics case against him, their cell-phone conversation was taped by Democratic loyalists using a police scanner. Cell-phone scramblers do exist today, but they tend to cost several hundreds of dollars. Chaotic masking might turn out to be a cheaper alternative for defeating casual eavesdroppers.

For military and financial applications, on the other hand, much stronger encryption is required. So far, chaos-based methods have proved disappointingly weak. Kevin Short, a mathematician at the University of New Hampshire, has shown how to break nearly every chaotic code proposed to date. When he unmasked the Lorenzian chaos of Cuomo and Oppenheim, his results set off a mini–arms race among nonlinear scientists, as researchers tried to develop ever more sophisticated schemes. But so far the codebreakers are winning.

One of the most promising developments comes from the 1998 work of Gregory VanWiggeren and Rajarshi Roy, physicists then working at the Georgia Institute of Technology. They gave the first experimental demonstration of chaotic communications using lasers and fiber optics, instead of electrical generators and wires. In their optical system, chaotic waves of light carried hidden messages from one laser to another at speeds of 150 million bits per second, thousands of times faster than the rates achieved electronically. And there's no theoretical barrier to even higher speeds.

Another advantage of communicating with chaotic lasers is that the chaos is much more complex, making it tougher to crack. The complexity is quantified by a number called the dimension of the strange attractor, which is a natural generalization of the ordinary concept of dimension. But unlike a straight line (which is one-dimensional) or a flat plane (which is two-dimensional), strange attractors typically have dimensions that are fractions. The Lorenz attractor, for example, is made of infinitely many two-dimensional sheets, which implies that it has an infinite surface area but no volume. Arcane as it may sound, it's more than a surface but less than a solid, and its dimension, accordingly, is greater than 2 but less than 3. For VanWiggeren and Roy's erbium-doped fiber lasers, the dimension of the strange attractor is unknown but it is almost certainly a fraction and, more important, it is huge. It seems to be at least 50, corresponding to an extremely wild form of chaos. It remains to be seen whether this new form of encoding will be more secure than its predecessors.

Leaving encryption aside, the more lasting legacy of synchronized chaos may be the way it has deepened our understanding of synchrony itself. From now on, sync will no longer be associated with rhythmicity alone, with loops and cycles and repetition. Synchronized chaos brings us face-to-face with a dazzling new kind of order in the universe, or at least one never recognized before: a form of temporal artistry that we once thought uniquely human. It exposes sync as even more pervasive, and even more subtle, than we ever suspected.

Eight

SYNC IN THREE DIMENSIONS

M Y FIRST ENCOUNTER WITH SYNC OCCURRED BY chance on a dismal day in Cambridge, England, in 1981. I was studying math there on a Marshall Scholarship after graduating from college, and feeling entirely displaced. The English girls never got my jokes, the brussels sprouts were gray, the drizzle was relentless, and the toilet paper was waxy. Even my coursework was drab: old-fashioned topics in classical physics, like the rotational dynamics of spinning tops. It was complicated stuff, and not inspiring.

Hoping to rekindle my academic passion, I walked across the street to Heffer's Bookstore to browse the books on biomathematics. (As a senior in college, I had written a thesis about the geometry of DNA, and that whole experience—doing original research with a world-class biochemist, using some of the math I was learning and applying it to an unsolved problem about chromosome structure—had been so thrilling that I was convinced I wanted to become a mathematical biologist.) As I scanned the shelves, with my head tilting sideways, one title popped out at me: *The Geometry of Biological Time.* Now that was a weird coincidence. My senior thesis on DNA had been subtitled "An Essay in Geometric Biology." I thought I had invented that odd juxtaposition, geometry next to biology. But the book's author, someone named Arthur T.

Winfree, from the biology department at Purdue University, had obviously connected them first.

The blurb on the back flap looked promising: "From cell division to heartbeat, clocklike rhythms pervade the activities of every living organism. The cycles of life are ultimately biochemical in mechanism but many of the principles that dominate their orchestration are essentially mathematical." I dipped into the table of contents. Right away I could see that this was the work of an unusual scientist. No, not just unusual. Arthur T. Winfree was breaking all the rules. Above all, he was playful. In a chapter about the mathematics of the menstrual cycle, he used data from his own mother. Other chapters had equally quirky elements in them: puns in their titles, personal stories from the author ("Nixon chose that week to invade Cambodia"), and I started to wonder if Winfree was for real. So I slid the book back into its place on the shelf, and left the store.

A few days later I felt myself being tugged back to Heffer's. Winfree's book was beckoning, and I had to look at it again. To check his credentials, I turned to the bibliography: 36 papers between 1967 and 1979, with several in the most prestigious journals, such as *Science, Nature,* and *Scientific American.* That should have been convincing enough, but for some reason I put the book back again, only to revisit it a few days later. Eventually it occurred to me that this was getting ridiculous—and God forbid that someone else might snatch the store's only copy. I surrendered and bought it.

Every day of reading the book was a new delight. Winfree's synthesis was brilliant and utterly original. Chapter by chapter, he built a mathematical framework that exposed an underlying unity in how various biological oscillations work. He applied his ideas to heart rhythms, brain waves, menstrual cycles, circadian rhythms, the cell division cycle, even waves in the gut. But his ideas went far beyond that. They made startling predictions that had kept turning out right in experiments. Some of them dealt with matters of life and death.

For the first time, I could sense my career path beginning to unfold. Excitedly I wrote to Winfree to ask for ideas about where to go to graduate school for mathematical biology. (I hadn't heard of any formal programs in it. The subject was too new, too much on the fringe.) Two weeks later, my pulse quickened when I picked up the mail and spotted the Purdue return address. Inside, scrawled in red Magic Marker on blue-lined school paper, with a few phrases connected by swooping arrows, was a reply from Winfree himself:

> Steven Strogatz:
> Well, of course you should come to me.

And after two pages of generous advice, he closed with:

> Do keep in touch: You sound interesting.
> Art Winfree

That was a dream come true. By then Winfree had become my hero. But he was in a biology department, and a graduate degree in biology was not in my plans—math was my subject. So how about a summer job with him? I sheepishly raised that possibility. Two weeks later, a reply arrived:

> 12-10-81
>
> 5 min after receiving yours of 12-1-81
>
> Dear Steven——
>
> This week a pile of $ fell on me so yes, I can provide a summer salary [. . .]
> There is plenty of space in my lab and 2 Apple computers w/ various wonderful
> attachments. [. . .] I will be working at topol. puzzles about 3-D twisted +
> knotted waves in Zhabotinsky' soup, + "moonlighting" applications to cardiac
> muscle (My Scient. Amer. article on sudden cardiac death will fill you in this
> spring.) I would be super-delighted to enlist your partnership in these
> endeavors. I think we could learn a lot together.
>
> I will not encourage [. . .] or [. . .] or anybody else to offer you a position
> until you decline this one. I hope you won't.
>
> Impulsively,
> Art Winfree

. . .

Winfree's research agenda, stated in idiosyncratic code in his letter to me, was ahead of what everyone else was thinking about. Of course he was well outside the mainstream of normal science, with its tendency toward narrow specialization and its emphasis on reductionism, drilling down to smaller and smaller units of inquiry—he wasn't thinking about single genes or quarks or neuron channels. But he was even outside the chaos revolution, which all of its practitioners felt was in the vanguard, but which was, in fact, already reaching maturity and about to give way to the next great trend: the study of nonlinear systems composed of enormous numbers of parts. Later christened as "complexity theory," this movement would come to seem like a natural outgrowth of chaos, in some ways its flip side—instead of focusing on the erratic behavior of small systems, complexity theorists were fascinated by the organized behavior of large ones. Winfree's earliest work on spontaneous synchronization of biological oscillators had already touched on that theme. By now it had matured in several ways.

For example, his letter mentioned his plans to work on "3-D twisted + knotted waves." The key phrase here is *3-D*. No one had ever looked into the behavior of self-sustained oscillators interacting in three-dimensional space. As we have seen earlier, when theorists first started analyzing the dynamics of oscillator populations, they ignored space altogether and concentrated on time alone, on rhythms in step, with no regard for how the oscillators were situated geographically. The breakthroughs of Wiener, Kuramoto, Peskin, and even Winfree himself had been restricted to the simplest possible case of all-to-all coupling, where each oscillator affects every other one equally. Global coupling was always recognized as nothing more than an expedient first step—it was the quickest way into the jungle of many-oscillator dynamics. There was no spatial structure to worry about; every oscillator is a neighbor to every other. Once that case was in hand, the next step up on the theoretical ladder was to consider oscillators arranged in a one-dimensional chain or ring. As you might expect, now something new can happen, something beyond pure synchrony: Waves of activity can propagate steadily from one oscillator to the next. In fact, in oscillator models with local coupling, waves turned out to be more common than sync. That makes intuitive sense from our own experience as fans at a football game: In a huge stadium, it's a lot easier to start "the wave" and keep it going than it would be to get the whole crowd standing up and sitting down simultaneously. When a few mathematicians tried to climb even higher, to two-dimensional sheets of oscillators, they had to hold on for dear life. The analysis became almost intractable. So when Winfree decided to keep climbing—to go three-dimensional—no one else followed.

The reason for thinking about such questions, of course, is that most real oscillators are coupled locally, not globally. The intestine is a long tube of oscillating nerve and muscle cells, segmented into rings that squeeze rhythmically, but choreographed so that waves of digestion travel in the right direction, from the stomach to the anus. Each ring of oscillatory tissue is coupled electrically to its nearest neighbors on either side, making the intestine effectively a one-dimensional chain of oscillators. The stomach is something like a two-dimensional bag of neuromuscular oscillators, in the sense that its cells churn rhythmically and interact mainly with their neighbors along the surface of the stomach wall. And the heart is a thick, three-dimensional collection of dictatorial oscillatory cells (the pacemaker cells in the sinoatrial node and their subordinates) and subservient "excitable" cells that obey their commands; if triggered by a strong enough electrical stimulus, they fire once and return to rest, awaiting the next triggering pulse. When the heart is functioning normally, the pacemaker generates a wave of electrical excitation that spreads along specialized conduction fibers to the pumping chambers (the ventricles), causing them to contract and pump blood to the rest of the body.

In pathological cases, however, the excitable cells can mutiny and sustain a wave of their own, a rotating electrical tornado that fends off the incoming signals from the pacemaker. Cardiologists had known for decades that such "rotating action potentials," or "circus movements," could lead to tachycardia (abnormally fast heartbeat) and then degenerate into the lethal arrhythmia called ventricular fibrillation, where the heart muscle writhes helplessly, twitching and quivering but not pumping any blood. Every year, hundreds of thousands of apparently healthy people—people with no prior history of heart disease—die suddenly when their hearts fall into this pernicious mode of organization. When Winfree mentioned "'moonlighting' applications to cardiac muscle" in his letter, he was referring to these strange electrical tornadoes. He wanted to find out why they start, how they behave, and what could be done about them. Once the basic science was understood, he hoped, it should enable the design of defibrillators that are gentler than today's crude devices, which burn the heart in order to save it.

In 1981, nonlinear dynamics had certainly not advanced to the stage where it could predict the behavior of such rotating waves in three dimensions. There was no hope of calculating their evolution in time, their lashing about, their swirling patterns of electrical turbulence. Even if the calculations were possible (assisted by a supercomputer, perhaps), any such attempt would be premature, since one wouldn't know how to interpret the findings. In fact, no one even knew what a mug shot of one of these shadowy villains might look like. (They'd never been seen directly by cardiologists.) So Winfree felt that the first step was to learn how to recognize them, to anticipate their features in his mind's eye; he would worry about their modus operandi later.

For the study of shapes in three dimensions, a coarser mathematics was needed, one that didn't care about time but only about space. When Winfree mentioned "topol. puzzles," he was referring to the branch of mathematics called topology, the study of continuous shape, a kind of generalized geometry where rigidity is replaced by elasticity. It's as if everything is made of rubber. Shapes can be continuously deformed, bent, or twisted, but not cut—that's never allowed. A square is topologically equivalent to a circle, because you can round off the corners. On the other hand, a circle is different from a figure eight, because there's no way to get rid of the crossing point without resorting to scissors. In that sense, topology is ideal for sorting shapes into broad classes, based on their pure connectivity. Winfree's plan was to use topology to classify the kinds of waves one might encounter in three-dimensional fields of excitable cells. Knowing what was possible, he'd know what to look for in later experiments and would have a hope of recognizing what would otherwise seem like bizarre, alien structures.

When I arrived at Winfree's lab on a muggy day in June 1982, he was engrossed in some paperwork, sitting alone at a lab bench with his shirt wide open. I was a little embarrassed by the informality—my dad had accompanied me on the cross-country drive from Connecticut to Indiana, and this was his first look at my new hero—but Winfree disarmed us with his unbuttoned friendliness. Soon my dad took his leave, and it was just Winfree and me alone in his lab, with its beakers and Bunsen burners and razor blades everywhere. (I later found out that razor blades were his tool of choice for cutting. He'd happily shout *"Zzzzupp!"* whenever he wielded one to slice a piece of wire or millipore filter paper.)

The lab was quiet. No grad students or postdocs. But I was prepared for that—in earlier correspondence, when I'd asked who else would be working with us, Winfree wrote back, "Now I could make up tales about the other students + co-workers. But truth to tell, I have none. Maybe I am away too much to form relationships, maybe I have body odor, dunno . . . but population density = 1 in my lab. You will be a singular event. Does that undermine your confidence?"

We had only three months to work together, so I needed to learn quickly. Winfree felt I should get my hands dirty: no math or computers for a while. My first project was an experiment on what Winfree called Zhabotinsky soup, a chemical reaction that supports waves of excitation remarkably like the electrical waves that trigger the heartbeat. But it's much simpler than a real heart—it's not even alive—and it has no muscles or motion of any kind. It's an idealized arena for exploring excitable wave propagation in its purest form. In that way, it plays the same role for heart waves that fruit flies play for genetics: a convenient simplification that captures the essence of more complicated phenomena.

Normally, the most amusing outcome you can hope for in a chemistry experiment is a puff of smoke or a noxious odor. In comparison, Zhabotinsky soup offers nonstop entertainment. When brewed according to its original recipe, it acts like a spontaneous oscillator, the chemical analog of pacemaker cells. It changes colors back and forth, rhythmically alternating between sky blue and rusty red dozens of times, before eventually relaxing to equilibrium about an hour later. At the molecular scale, the performance would appear even more impressive, if only we could see it: trillions of coupled oscillators, hoofing in perfect sync, the largest line dance ever assembled.

In its new, more subtle recipe, the reaction is excitable. At first it looks disappointingly inert. The oscillations are gone. But if you pour a thin layer of the red soup into a petri dish and then prick it with a silver wire or a hot needle, it suddenly launches a blue circular wave that expands and spreads like a grassfire. This is a chemical wave, a pulse of propagating excitation in which the reaction switches from a reduced state to an oxidized one. Once the wave has passed, the reaction reverts to quiescence and turns red again, just as grass eventually grows back after a grass fire. (This analogy is not perfect, however. The chemicals recover more rapidly than the prairie; a second wave can follow right behind.)

Chemical waves are completely different from the waves studied in traditional physics courses, like sound waves or the ripples on a pond. When a chemical wave spreads by diffusion, the surface of the liquid does not bob up and down. It remains motionless. What moves is a pattern of excitation, a kind of chemical contagion. Nor do these waves weaken like sound or ripples as they travel away from their origin. Each patch of the medium provides a fresh source of energy that refuels the wave, preventing it from damping out.

Now suppose you detonate two chemical waves at two different points in the petri dish. The blue circles expand and creep toward each other. When they collide, they do not interpenetrate or add up: They annihilate. And they do so for the same reason that onrushing grass fires snuff each other out: Neither can burn through the other's ashes. In this metaphor, the ashes correspond to a region of exhaustion, a refractory zone in the wake of the wave. The chemical medium needs time to recover before it can become excited again.

In many ways, this chemical medium behaves like the human sexual response. Sexual arousal and recovery depend on the properties of nerve tissue, which, like Zhabotinsky soup, belongs to a general class of systems called excitable media. A neuron has three states: quiescent, excited, and refractory.

Normally a neuron is quiescent. With inadequate stimulation, it shows little response and returns to rest. But a sufficiently provocative stimulus will excite the neuron and cause it to fire. Next it becomes refractory (incapable of being excited for a while) and finally returns to quiescence. The parallels with chemical waves extend to action potentials, the electrical waves that propagate along nerve axons. They too travel without attenuation, and when two of them collide, they annihilate each other. In fact, all of these statements are equally true of electrical waves in another excitable medium: the heart. That's the beauty of this abstraction—the qualitative properties of one excitable medium hold for them all. They can all be studied in one stroke. The family resemblance among Zhabotinsky soup, nerve tissue, and heart muscle persists right on down to the structure of the mathematical equations that govern their nonlinear dynamics. The analogy runs deep.

But Zhabotinsky soup offers a number of advantages, especially for a beginning experimenter. No animals need to be sacrificed. There's no confusing anatomy, like the intricate tangle of neural networks or the twisted-fiber architecture of the heart muscle. Best of all, the waves are visible to the naked eye and they move slowly, so there's no need for any elaborate recording equipment. In contrast, the visualization of waves on the heart remains a formidable technical challenge to this day, even for labs with huge budgets, requiring voltage-sensitive dyes, multielectrode arrays, and other state-of-the-art technology.

With the help of Zhabotinsky soup, scientists have begun to unravel the secrets of wave propagation in excitable media. In particular, it was in Zhabotinsky soup that a new kind of wave was discovered: a rotating, self-sustaining wave shaped like a spiral. Although its geometry is graceful, its consequences are destructive. Rotating spiral waves on the heart are the culprits behind tachycardia and, in the worst case, ventricular fibrillation followed by sudden cardiac death.

The discovery of Zhabotinsky soup and its remarkable spiral waves is a tale of dogma, disappointment, and ultimate vindication. Of course, Zhabotinsky soup is not its real name—that's just what Winfree always called it. Today it's known as the BZ reaction, for Belousov and Zhabotinsky, the Russian scientists who invented it and refined it, respectively.

In the early 1950s Boris Belousov was trying to create a test-tube caricature of the Krebs cycle, a metabolic process that occurs in living cells. When he mixed citric acid and bromate ions in a solution of sulfuric acid in the presence of a cerium catalyst, he observed to his astonishment that the mixture became yellow, then faded to colorlessness after about a minute, then returned to yellow a minute later, then became colorless again, and continued to oscillate dozens of times before finally reaching equilibrium after about an hour.

Nowadays it comes as no surprise that chemical reactions can oscillate spontaneously; such reactions have become a standard demonstration in chemistry classes. But in Belousov's day, his discovery was so radical that no one would believe it. It was thought that all solutions of chemical reagents must go monotonically to equilibrium, because of the laws of thermodynamics. Journal after journal brushed off Belousov's paper. One editor even salted his rejection letter with a snide remark about Belousov's "supposedly discovered discovery."

Dejected, Belousov resolved never to share his breakthrough with his chemist colleagues. He did publish a brief abstract in the obscure proceedings of a Russian medical meeting, but hardly anyone noticed it until years later. Nevertheless, rumors about his amazing reaction circulated among Moscow chemists in the late 1950s, and in 1961 a graduate student named Anatol Zhabotinsky was assigned by his adviser to look into it. Zhabotinsky confirmed that Belousov had been right all along, and brought this work to light at an international conference in Prague in 1968, one of the rare occasions that Western and Soviet scientists were allowed to meet. At that time there was fervent interest in biological and biochemical oscillations, and the BZ reaction was seen as a manageable model of those more complex systems.

The analogy to biology turned out to be surprisingly close. In early 1970, A. N. Zaikin and Zhabotinsky found propagating waves of excitation in thin, unstirred layers of BZ reaction. The waves resembled concentric circles, and they annihilated upon collision, just like electrical waves in neural or cardiac tissue. They even seemed to emanate from something analogous to pacemakers, randomly scattered points that belched waves spontaneously.

After learning of this work, Winfree wrote to Zhabotinsky (whom he'd met two years earlier as a fellow grad student at the Prague conference) to ask whether he'd ever seen any other wave patterns besides concentric rings. Winfree had observed spiral waves in his own lab experiments on a certain kind of fungus, but that was a far more complex system composed of living creatures with circadian clocks. He wondered if spirals could also occur in Zhabotinsky's much simpler chemical system. He doubted it on mathematical grounds; he thought he could prove that the waves had to be closed rings. But still no reply from Zhabotinsky. The mail from the Soviet Union was maddeningly slow in those days, especially between scientists (national security agencies at both ends were probably busy steaming open the envelopes). Winfree couldn't bear the suspense. He concocted Zaikin and Zhabotinsky's recipe for himself, and sure enough, spirals popped up everywhere. Winfree had no way of knowing it, but Zhabotinsky had also seen them in his 1970 thesis work, and Valentin Krinsky in Puschino had anticipated them in any excitable medium, heart muscle included. Spiral waves are now recognized to be a pervasive feature of all chemical, biological, and physical excitable media.

Boris Belousov would be pleased to see what he started.

In 1980, he, Zhabotinsky, and three other scientists were awarded the Lenin Prize, the Soviet Union's highest medal, for their pioneering work on oscillating reactions. But it wasn't much consolation—Belousov had died 10 years earlier.

The most striking thing about spiral waves is that they seem to be alive. They're self-sustaining. They don't need pacemakers: A spiral wave is its own pacemaker. If you watch one in a thin layer of excitable BZ reaction, it looks like a perpetual pinwheel, chasing its tail and regenerating itself endlessly.

In a way, the rotation is merely incidental. More fundamentally, the wave is propagating, advancing perpendicular to itself at each point along the wavefront. The confusion occurs because of a quirk about spiral geometry: Propagation looks like rotation. (Think of the optical illusion seen on old barbershop poles. The helix painted on the rotating pole seems to be propagating upward. But of course it's not moving up at all; it's merely turning along with the pole. Here rotation masquerades as propagation—the converse of the same effect in spiral waves.)

Nevertheless, there is a sense in which the rotation of a spiral wave is real. Each point in the surrounding medium oscillates periodically; it's re-excited whenever the wave passes through. So every point in the petri dish cycles through the familiar stages of excitation, refractoriness, quiescence, and then re-excitation. What's new here is that the spiral wave has created an oscillation that's structured in space as well as time. Instead of lockstep synchrony—the spatial uniformity that Belousov saw in his earliest experiments, where the whole beaker changed color at once—the oscillation is now like "the wave" initiated by the fans at a football game, which circulates around the stadium as people stand up and sit down at just the right time.

For an even closer analogy, imagine a ring of a thousand dominoes carefully arranged on the floor. Suppose that we have enlisted the help of a speedy assistant who agrees to reset any domino immediately after it has fallen. We tip the first domino, and a wave of toppling begins to propagate around the ring. The assistant follows close behind, furiously resetting the dominoes. Here a tipping domino corresponds to the excited state, a fallen one is refractory, and an upright one is quiescent. Such a wave will circulate endlessly, or until the assistant collapses.

A biological version of the same experiment was done by the physiologist A. G. Mayer in 1906 with the help of a jellyfish. He fashioned a ring of neuromuscular tissue from the rim of the jellyfish's umbrella-shaped dome, and then electrically stimulated it at one point, taking care to allow wave propagation in one direction only. The neural impulse circulated for six days, executing about half a million cycles.

So it's clear that waves can circulate persistently around one-dimensional loops of excitable media. But there's a problem with extending the same ideas to two dimensions, the important case for spiral waves. In the discussion above, we implicitly assumed that the medium had recovered from its refractory period by the time the wave returned. That's a valid assumption if the loop is big enough or if the wave speed is slow enough. But near the center of a spiral wave, this assumption breaks down: The loop traversed by the excitation has become too small.

The upshot is that the central core of the spiral does not oscillate like the rest of the medium. It doesn't display rhythmic variations in color, or peaks and troughs in light intensity, or any other sign of oscillation. The cycle amplitude drops to zero. Such a point is called a phase singularity, meaning that the phase of the surrounding oscillation cannot be sensibly defined there. Phase becomes ambiguous. This puzzling situation is analogous to what happens at the North and South Poles. At those singular points on the surface of the Earth, all the time zones converge and the cycle of day and night breaks down. The sun neither rises nor sets; it merely circles along the horizon. So it's senseless to ask what time it is at the poles. It's all times, and no times.

But for a spiral wave, the phase singularity is more than a point of remote geographical interest. It's the engine that drives the wave. Amazingly, as long as the core is intact, the entire spiral wave can regenerate itself, no matter what damage is done to its outer arms. And spiral waves are tough to eradicate for another reason: They emit waves almost as fast as the medium will allow. So they are able to fend off other incoming waves, such as the concentric circles launched by distant pacemakers. The encroaching waves are annihilated in collisions with the spiral arms. They can make no headway. On the contrary, the faster spiral waves inexorably advance on the slower pacemakers, usurping their territory and eventually snuffing them out. That's why, in the long run, a dish of BZ reaction always looks like a paisley pattern filled with spirals, with no circular waves in sight. Only one spiral can resist another.

Here we see a case of spontaneous order, pure and simple. Start with a soup of chemicals that happens to be excitable. Then touch it with a silver wire and slosh it around to set up a random pattern of excitation. No structure, just a

mess, and yet out of it emerges a paisley. There's nothing mystical about it. The pattern follows from the laws of excitable media, and those laws in turn come from nonlinear dynamics.

By playing with Zhabotinsky soup in Winfree's lab for a few days, I picked up the basic facts about spiral waves. For my next task, Winfree suggested that I try to reproduce an experiment about a new kind of spiral wave that had recently been submitted to *Nature*. After a few weeks of failure, it became apparent to Winfree that I was an experimental clod (no news to me of course—such ineptitude takes years of refining).

Fortunately, Winfree's main goal for the summer was in a completely different direction. As he mentioned in his letter, he wanted to work on "puzzles about 3-D twisted + knotted waves in Zhabotinsky' soup." The questions were: What are the three-dimensional generalizations of spiral waves? What do they look like? Can we visualize them? And what are the mathematical rules governing their allowed shapes?

He'd already made a good start. Soon after his discovery of two-dimensional spiral waves in 1970, he imagined what would happen if he took a thin layer of BZ reaction harboring a flat spiral and then progressively deepened the layer. Like a bas-relief, the spiral would rise into the third dimension and sweep out a continuous stack of spirals—a surface shaped like a scroll.

Meanwhile, the singular point at the core would elongate to a singular filament at the edge of the scroll. And just as a spiral wave rotates around its core, a scroll wave must rotate around its filament.

A rotating scroll wave: Science had never seen anything like it. It was not easy to find analogies. A scroll wave is a chemical tornado. Yes, except that the liquid remains motionless. What moves is a wave of chemical activity, a three-dimensional vortex of spreading excitation. Furthermore, tornadoes reach from

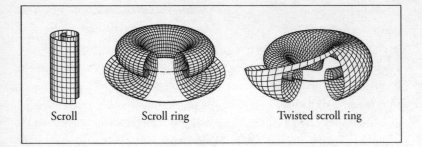

Scroll   Scroll ring   Twisted scroll ring

the clouds to the ground—but where do scroll waves end? Winfree convinced himself that they could not just stop somewhere in the middle of the liquid. They would either have to terminate on the boundary—the walls of the beaker, or the air-liquid interface at the top—or maybe they didn't have to end at all. In other words, a scroll wave might bite its own tail and close on itself. Instead of a tornado, it would be more like a smoke ring.

This image captivated Winfree. Do such "scroll rings" really exist? In 1973 he proved that they did. His experiment was ingenious. Instead of the usual beaker full of liquid BZ reaction, he prepared a tall stack of porous nitrocellulose filter papers saturated with the same chemicals. After setting up what he thought were the right conditions to conjure a scroll ring, he let the reaction proceed and then suddenly fixed it chemically, capturing the telltale pattern in a state of suspended animation. To examine the specimen, he sliced the stack into thin layers, like a microscopist preparing slides of an exotic organism, and then reconstructed it slice by slice on sheets of nonreflecting glass. The specimen turned out to be just as expected: a doughnut-shaped wave with a spiral cross section.

But then Winfree wondered if other kinds of scroll rings could exist. Could scroll waves be twisted through a whole number of turns before closing? Belts can twist in this way: Why not scroll rings? Or could they be tied in knots? Could rings link through one another, like bracelets or chain mail? As he considered the specter of an infinite zoo of scroll rings, linked and twisted and knotted in diverse ways, Winfree soon discovered that one of the hypothetical beasts was forbidden.

Using a theorem from topology, Winfree proved that a twisted scroll ring was impossible, at least as a solitary entity. Its structure was self-contradictory. If the ring were twisted, it automatically had to be threaded by another singular filament, and that meant that the original ring was not alone after all. The topological theorem had unveiled a second scroll, unforeseen but guaranteed to be linked through the first. With further effort, Winfree could show that although an isolated twisted scroll ring was forbidden, a mutually linked pair was not. It seemed to be a perfectly viable structure.

The implication was tantalizing: Scroll ring geometry was lawful. Some configurations were admissible, while others were not. There were rules waiting to be discovered.

The first order of business was to picture what twisted scroll rings would look like. Winfree's abstract topological argument implied that a twisted scroll ring had to be threaded by another singular filament, but neither he nor I could picture how the entire structure—the twisted scroll plus its additional threading singularity—would fit together globally. In fact, when Winfree had once tried to hand sketch it years earlier, he accidentally produced a nonsense picture in the style of Escher, like the one in which zombies are climbing four flights of stairs that impossibly lead back to the bottom landing.

But now everything would be different. This was the modern era—1982—and we had Apple computers. The computer could draw the surface for us; we just had to tell it what to draw. My job was to write a computer program that would calculate the surface by brute force. The idea was simple: The twisted scroll is nothing more than a circle of spiral ribs, each of which is cocked slightly compared to its neighbors. So I told the computer to calculate a bunch of points on a spiral, then copy and advance that entire spiral one notch around the circle, and twist it by one notch at the same time. Do this over and over, until the spiral returns to the starting position, having made one orbit around the circle and one full twist. The only tricky thing was, how long should each spiral rib be? That is, how many turns should it have? Here chemistry gives the answer: The spiral wave keeps going until it bangs into another one. Portions of the colliding spirals beyond that should be erased, because they would have annihilated each other (as colliding waves do in excitable media).

As requested, the Apple IIe spat out a table of a few hundred numbers, representing a meshwork of points on the twisted scroll surface. Now all we needed to do was run those numbers through a graphics program, and it would finally unveil the twisted scroll ring. I revved up the software that Winfree had invested in—*Bill Budge 3D Graphics System*—and we held our breath. Hmmm. The pictures turned out to be too coarse—not enough mesh points. Unfortunately, Bill Budge's system couldn't take anymore—it was already groaning under our demands. Our last resort was to connect the dots by hand. We printed out the clunky pictures, and used colored pencils to embellish the hard copy, hoping to see something stunning. No luck. The unveiling would have to wait.

Meanwhile Winfree and I started on the more theoretical questions, searching for the rules of scroll-wave topology. With no clear direction to go, we felt like we needed better intuition. Winfree kept big blobs of red and green dental wax in his lab, along with orange molding clay and an endless supply of pipe cleaners. All of these were indispensable for making sculptures of knots and links and twisted surfaces.

That was how we worked. While I sat at the computer or the lab bench, molding dental wax and trying to visualize shapes never seen before, he would draw scroll-wave pictures on an artist's sketchpad, always in Magic Marker, and then *zzzzupp* the good ideas with a razor and tape them into his lab notebook. Hours would pass. Occasionally one of us would interrupt the silence with what felt like an insight. Then we would struggle to explain the vision, and clarify it, and check it for sensibility, always grasping for words because three-dimensional geometry is so elusive and difficult to convey. But eventually we always understood each other, and together we started to hammer the ideas into the glimmerings of a theory. These mathematical conversations were intense but exhilarating. For me, it felt like having a second brain, only a much better one. This went on all day long, day after day. Usually we ate lunch together, and on sunny days we'd sit by the pool at his apartment, while he sketched on his pad and I pictured surfaces in my head. By ten at night, one of us usually had a headache and we'd quit.

By August, we had figured out the rules for all possible configurations of linked and twisted rings. But knots were hard. We didn't know any rules for them. So we started with the simplest case: a single scroll ring, with a trefoil knot tied in it. (To make a trefoil, take a shoelace, tie an overhand knot in it as if beginning to tie your shoe, and then fuse the tips together. The resulting curve is a knotted loop that looks something like the silhouette of a three-leaf clover.)

We wondered whether a trefoil-shaped scroll ring could make mathematical and chemical sense. If it were sitting in a beaker of BZ reaction, would it always have to be linked by other rings, or could it exist on its own, if it were also twisted in the right way? And if so, how much twist was right? What would the waves emanating from it look like?

To make these abstractions more palpable, I rolled some dental wax into long, stringy pieces, and then bent them and squished their ends together until they looked like a trefoil. That was supposed to be the singular filament, the source and inner edge of the scroll wave. Next came the challenge of making a wax model of the scroll-wave surface itself. If the singular filament is like the long, thin wooden dowel of a scroll, the wave is like the parchment that unrolls from it. It's a surface that begins and ends at the dowels, while curling tightly around them at the same time. Fortunately, the curling was inessential, in a mathematical sense: It could always be removed by pulling the scroll wave taut (imagine the wave is made of spandex). What's crucial about the scroll wave is that it begins and ends on the filament. It's a surface that has no other boundary. With another color of wax, I began developing the wave surface, one patch at a time, always starting along the filament and working my way in, until all the patches merged into one continuous sheet.

The next question was, Does that sheet have one side or two? That might sound crazy: What is a one-sided surface? The most famous example is a Möbius strip, a ribbon of paper that is given half a twist and then closed to make a ring. If you start with your finger somewhere, and then trace around the loop, your finger eventually comes back on the other side of the paper (though that's the wrong way to say it—there is no "other" side; the front and back are the same). In that sense, the Möbius strip has only one side.

If my wax surfaces were like this, it would be bad. Chemistry dictates that the scroll wave has to be a two-sided surface, because of a basic fact about excitable media: Waves propagate perpendicular to themselves, burning into quiescent territory and leaving refractory ashes behind. That means that the wave has both a front and a back, but a Möbius strip doesn't. Or to say it another way, imagine that you paint one side of the Möbius strip red—the side that's supposed to burn forward—and then paint the other side black—the side where the ashes are. But they're the same side, so you'll end up painting black on top of red. The whole notion of forward propagation makes no sense if the wave is one-sided.

There are various ways of drawing a trefoil. Curiously, some lead to one-sided surfaces (and are therefore forbidden) while others give the desired two-sided surfaces, providing candidates for the shape of the wavefront. After a bit of playing around, I realized that all acceptable surfaces were topologically equivalent; with the appropriate bending and stretching, each could be deformed continuously into any of the others. So there was only one right answer, and here it was. This is what the scroll-wave surface for a trefoil would have to look like.

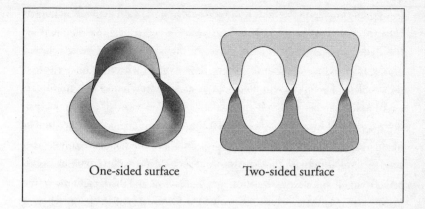

One-sided surface   Two-sided surface

The remaining question was whether the resulting scroll would be twisted, and if so, to what extent. To measure the twist experimentally, I laid a piece of thread along the wax surface, always running parallel to its outer edge but just a millimeter inside it, and continued it all the way around the surface until it closed into a loop. That loop also formed a trefoil knot, just like the original filament, and together they defined the two edges of an imaginary ribbon.

This ribbon reminded me of something in my senior thesis in college, which had dealt with the topology of supercoiled DNA molecules. A key concept there was a mathematical quantity called the linking number of DNA, which, roughly speaking, measures how many times one strand of DNA winds around the other, above and beyond the winding implied by the double helix itself. It depends on both the twist in the DNA and its three-dimensional path through space. Now, for scroll waves, the linking number of the ribbon would contain all the important information about the twist of the wave, as well as the shape of its knotted filament. When I calculated the linking number, it came out to be zero. Beautiful—it was that simple. Trefoil-shaped scroll waves can exist, and they always come with zero twist. Later we proved that the same must be true for any knot, not just a trefoil.

When the summer ended, I moved to Boston to begin graduate school at Harvard, but Winfree and I kept in touch. We had papers to write, and we still had two lingering puzzles to solve. In the winter of that year, I visited him at his parents' home in Longboat Key, Florida, where we finally solved the problem of scroll-wave topology in its most general form. We proved that an arbitrary number of scroll rings could be linked, twisted, and knotted in diverse ways, as long as they satisfied a single equation: The linking number of each ring's ribbon, plus all its mutual linkings with the other rings, must sum to zero. Otherwise the structure was forbidden. With tongue in cheek, we called this the exclusion principle, by analogy with the Pauli exclusion principle in chemistry, which constrains the atomic structure of the elements and gives rise to the patterns in the periodic table. For us, the "elements" were the allowed configurations of scroll rings and knots, ranked in order of increasing complexity. "Hydrogen" was a single scroll ring, with no knots or twists in it. "Helium" was two rings, linked through each other and twisted once.

A few months later, we spent the summer at Los Alamos National Laboratory, working on the world's fastest supercomputer. (It was a Cray-1, but the local bomb makers had a more ominous name for it: the "X machine.") With the help of Mel Prueitt, the resident expert on computer graphics, we finally produced pictures of a twisted scroll ring that clearly unveiled the gnarled singularity that, on abstract mathematical grounds, we knew must thread through its center. Winfree and I both gasped when we saw it. It was like finally meeting a beloved pen pal from another country, whose face we had only dimly imagined.

In the 20 years since then, there's been an explosion of interest in spiral and scroll waves. Chemists have made much more careful measurements of the BZ reaction, using computer-aided video recording, and they've discovered that spirals don't always pivot about one point—they often meander. The inner tip of a spiral wave can rotate in circles, or trace out flower patterns, or even wander chaotically. Mathematicians have jumped on those results, eagerly explaining them as instabilities stemming from nonlinear dynamics.

In all of this, the Holy Grail remains cardiac arrhythmias. A number of cardiologists and physiologists have confirmed experimentally that spiral and scroll waves can cause tachycardia, though the route to ventricular fibrillation remains controversial. The most likely suspects are a meandering spiral wave, the disintegration of one spiral into many, and the thrashing instabilities of a three-dimensional scroll wave. Several teams of cardiologists and mathematicians are working intensely on this problem, and the true culprit could be identified soon.

During all this time, Winfree has relentlessly pursued scroll waves and their possible role in cardiac arrhythmias. He remains consumed with visions of knots and links; but now he is concerned with their dynamics, no longer the frozen geometry that we explored together. Equipped with the immense power of today's supercomputers, he and his students have simulated how linked and knotted scroll waves would move. Their filaments flail about, writhing violently, as waves from part of one filament slap against another. Yet many of these structures turn out to be remarkably stable; they do not spontaneously untie themselves. In that sense, they are fundamental, like the elementary particles in quantum physics. They are the basic localized solutions of the field equations for excitable media. They *have* to be important. That's why Winfree will never give up on them.

He has also sought (but not yet found) a simple law that might explain how the filaments slither and writhe. Even if there were an elegant answer, no one knows whether it would matter for arrhythmias: In cardiac muscle so far, only the most elementary scroll wave, a straight scroll with no knots or links, has been found. Undaunted, Winfree has returned to the lab bench and invented a new kind of optical tomography for the BZ reaction, in hopes of capturing a snapshot of his elusive particles. He has been showered with well-deserved recognition: a MacArthur genius award in 1984, the Einthoven Award in cardiology in 1989, and the Norbert Wiener Prize in applied mathematics in 2000. His son Erik, who was a prepubescent computer whiz when I knew him, recently won a MacArthur prize too—the first pair of father and son winners.

In terms of his contribution to coupled oscillators and sync, Winfree taught us what wonders exist when oscillators are allowed to mingle in space, how they organize themselves around points of timelessness, spawning spirals in two dimensions and scroll waves in three. In the years to come, scientists would begin exploring an even more general form of connectivity, with oscillators coupled not to their neighbors in ordinary space, but to their neighbors in a mysterious and powerful kind of network—the kind that connects us all by just six degrees of separation.

Nine

SMALL-WORLD NETWORKS

IN JOHN GUARE'S 1990 PLAY *SIX DEGREES OF SEPARATION*, a character named Ouisa ruminates about the mystery of life in a small world:

> I read somewhere that everybody on this planet is separated by only six other people. Six degrees of separation. Between us and everybody else on this planet. The president of the United States. A gondolier in Venice. Fill in the names. I find that A) tremendously comforting that we're so close and B) like Chinese water torture that we're so close. Because you have to find the right six people to make the connection. It's not just big names. It's *anyone*. A native in a rain forest. A Tierra del Fuegan. An Eskimo. I am bound to everyone on this planet by a trail of six people. It's a profound thought. . . . How every person is a new door, opening up into other worlds.

A few years later, on a snowy winter afternoon in Reading, Pennsylvania, three inebriated fraternity brothers at Albright College came to a similarly cosmic conclusion: Every American movie actor can be connected to Kevin Bacon (the pug-nosed dancing rebel in *Footloose* and star of the cult favorite *Tremors,* a film about giant carnivorous worms) in four steps or less. Charlie Chaplin, for instance, has a Bacon number of 3. He was in *A Countess from Hong Kong* with Marlon Brando, who was in *Apocalypse Now* with Laurence Fishburne, who was in *Quicksilver* with Kevin Bacon. Suspecting they were on to something big, the frat boys contacted the *Jon Stewart Show,* then a late-night talk show on MTV. They were invited to play a round on the air, after which the game spread over the Internet and triggered a nationwide craze among college students. The Oracle of Bacon—a Web site that automatically computes the shortest possible chain of costars between Kevin Bacon and any other film actor—was chosen by *Time* magazine as one of the top 10 Web sites of 1996. At the height of the craze it was receiving 20,000 hits a day.

Other parlor games soon followed. In 1999, "Six degrees of Marlon Brando" broke out as a fad in Germany, as readers of *Die Zeit* tried to link a falafel vendor in Berlin with his favorite actor through the shortest possible chain of acquaintances. And during the height of the Lewinsky scandal, the *New York Times* printed a diagram of the famous people within "six degrees of Monica," from Bill Clinton and Saddam Hussein to O.J. Simpson and, of course, Kevin Bacon.

Silly as all this seems, there is something serious going on here. As a society, we have become obsessed with connectedness. We are struggling to make sense of the complex networks that have recently infiltrated our lives, networks whose reach is immense, whose structure we can only dimly perceive, and whose functioning bewilders us. We're confused about the consequences of globalization, disoriented by the Web, worried about contagion in the financial markets, and terrified of al Qaeda. Sometimes the fears prove unfounded— Y2K never set off the catastrophic ripple effect that pessimists predicted. But on August 10, 1996, a fault in two electrical power lines in Oregon led, through a cascading series of failures, to blackouts in 11 states and two Canadian provinces, leaving about 7 million customers without power for up to 16 hours. The Love Bug worm, one of the worst computer attacks to date, propagated across the Internet on May 4, 2000, and inflicted billions of dollars' worth of damage worldwide.

Science itself reflects the network zeitgeist. For example, with the completion of the human genome project, the focus of molecular biology has shifted from the discovery of new genes to the analysis of gene networks. Traditionally, the genome has been viewed as a blueprint for the construction of proteins, which in turn act as the building blocks for the cellular structures and molecular machines essential to life. But today we see that metaphor as too static, too linear, a vestige of the assembly-line mentality of an earlier era. Some of the most important genes (the so-called regulatory genes) code for proteins that alter the activity of other genes, turning them on or off, forming circuits and feedback loops. The genome starts to seem less like a blueprint and more like a computer. The functioning of this computer—and its malfunctioning when cells turn cancerous—will not be deciphered until we understand the logic of gene networks.

Similarly, throughout the rest of science, researchers are only now beginning to unravel the structure of complex networks, from the nervous systems of simple organisms to the overlapping boards of directors of the largest companies in the United States. The size of these networks is often daunting: 30,000 genes in the genome, millions of species in the terrestrial ecosystem, billions of people on Earth, someday 10 billion pages on the Web. But the problem is knottier than that. Even if we were given the complete wiring diagram for any of these systems—a list of all the nodes (genes, species, people) and the connections between them—we wouldn't know what to compute. The mass of data would be overwhelming. Until we know what we're looking for, the secrets of complex networks will remain elusive.

What we need now are ideas: simple, organizing principles to guide us through the morass of data. If history is any guide, the most penetrating ideas will come from mathematics. By its very nature, the mathematical study of networks transcends the usual boundaries between disciplines. Network theory is concerned with the relationships between individuals, the patterns of interactions. The precise nature of the individuals is downplayed, or even suppressed, in hopes of uncovering deeper laws. A network theorist will look at any system of interlinked components and see an abstract pattern of dots connected by lines. It's the pattern that matters, the architecture of relationships, not the identities of the dots themselves. Viewed from these lofty heights, many networks, seemingly unrelated, begin to look the same.

In 1998, my former student Duncan Watts and I published the first comparative study of complex networks from this point of view. Our analysis revealed that whether the nodes in the network are neurons or computers, people or power plants, everyone is connected to everyone else by a short chain of intermediaries. In other words, the "small world" phenomenon is much more than a curiosity of human social life: It's a unifying feature of diverse networks found in nature and technology. Since then, we and many other scientists have begun to explore the implications of small-world connectivity for the spread of infectious disease, the resilience of the Internet, the robustness of ecosystems, and a host of other phenomena.

The study of complex networks is only the next logical step in a larger journey, the quest for a science of spontaneous order. The quest thus far has taken us from the most primitive form of coordinated behavior—a pair of identical rhythms in sync—through ever more intricate choreographies in time and space: from two oscillators to many, from identical oscillators to diverse ones, from rhythms to chaos, from global coupling to local interactions in space. The next step is to move to more general kinds of connectivity, where neighbors are defined in an abstract sense that need not be geographical. Just as the spatial coupling between nonlinear systems spawned a new form of collective behavior—self-sustaining spiral and scroll waves—that couldn't occur in simpler geometries, complex networks give rise to even richer forms of self-organization. In fact, complex networks are the natural setting for the most mysterious forms of group behavior facing science today. If the day should ever come that we understand how life emerges from a dance of lifeless chemicals, or how consciousness arises from billions of unconscious neurons, that understanding will surely rest on a deep theory of complex networks. At the moment, such a theory is almost inconceivable. But at least we know how to begin. We need to master the principles of network architecture, to learn how nature weaves her intricate webs. Fittingly, our first excursion into this territory was by way of sync, on a detour from what was supposed to be a study of crickets chirping in unison.

When I started teaching at Cornell in the fall of 1994, one of my first chores was to administer a rite of passage known as the qualifying exam. Four professors sit side by side in an otherwise empty classroom, while a graduate student stands alone at the blackboard, protected only by a piece of chalk. For a half hour, we grill him with math questions. He's supposed to think on his feet, working out the answers in front of us. If he seems to be handling a problem too comfortably, we stop him and move on, escalating the difficulty, probing for soft spots.

I was assigned to the applied math portion of the exam. We were scheduled to test four or five students, one of whom was Duncan Watts, a six-foot-two-inch Australian with a confident smile and the physique of a Green Beret. He'd come to Cornell because of his fascination with chaos theory. Back home he'd majored in physics, where he was one of the top students at the Defence Force Academy, and a finalist for a Rhodes Scholarship.

The chair of the session nodded in my direction. "Professor Strogatz will ask the first question." I asked Duncan to solve Laplace's equation in a crescent-shaped region, using the method of conformal mapping. The other professors glared at me. Apparently that was not a topic that the students were expected to have studied (which, being a newcomer, I had no way of knowing). Duncan sputtered for a few seconds, saying something about how he'd studied conformal mapping in college, though he hadn't thought about it for a while. Realizing my gaffe, I offered to ask a different question, but one of my colleagues seemed to relish the rise in room temperature and said no, let's see what he does with this.

Step by step, Duncan felt his way through the problem, clearly not remembering the standard way to solve it, but somehow finding a path to the right answer, almost as if by sheer determination to hang on. He may have been flustered, but he never showed it. In fact, he gave the odd impression of having fun.

That reaction made sense to me a few months later, when I noticed a photograph of him posted on his office door, showing him hanging by his fingertips on the face of Point Perpendicular, a sea cliff in Australia that drops 70 meters straight down to the ocean. I knew I'd just found my next Ph.D. student.

We began casting around for a suitable thesis project for him. Maybe a problem about using chaotic lasers for private communications, or one about oscillations in the vessels of the lymph system. But neither lasers nor lymph really moved us. After half a year of indecisiveness we were both feeling frustrated.

One day in the spring of 1995, I gave a lecture on firefly synchronization at the department of neurobiology and behavior, where my colleague Ron Hoy and his students work on cricket communication. I emphasized how little contact there'd been so far between the theory of synchronization and any real biological examples, and wondered if they could help, maybe by setting up some experiments on the collective behavior of crickets. One of the postdoctoral fellows, Tim Forrest, became excited; he had been a math major in college and was now an expert in bioacoustics. Yes, he said, he'd love to investigate how male crickets manage to chirp together in vast choruses, in their nightly effort to court females. He offered to catch some "animals," as he called them, and volunteered to devise a set of experiments that would allow us to test our mathematical models, or maybe even point us toward new ones.

Duncan liked the sound of this project and began meeting with Tim every few days. Meanwhile, we fantasized about the experiments we wanted to try. One dream was to measure the songs of all the crickets simultaneously, to track their second-by-second progress toward sync: something that had never been done for fireflies or circadian clock cells or any other collection of biological oscillators. Another was to test for a phase transition, long predicted by the models of Winfree and Kuramoto, but never tested empirically. Our plan here was to alter the coupling between the crickets systematically. At low coupling, when they can hardly hear one another, the difference in their natural chirp rates should prevent them from synchronizing. Like the runners on the track who can't stay together because they have too wide a range of natural abilities, the fast crickets should drift away from the slower ones if the coupling is too weak. The population would lapse into cacophony. On the other hand, if we could gradually increase the crickets' mutual influence (by making their chirps louder or longer, or by somehow making the crickets more sensitive), the Winfree/Kuramoto theory implied we should see a critical value of coupling, an abrupt transition where many crickets would suddenly begin trilling in unison.

Even if we didn't find a phase transition, we hoped to document how mutual synchronization breaks out in a real population. Earlier experiments by one of Tim's former advisers had shown that an individual cricket adjusts to the calls of others. Thus, if it hears a call just before it was planning to chirp, it shifts its neural clock ahead by the necessary correction. Or if it hears a chirp just after one of its own, suggesting that its behavior was premature, its nervous system automatically corrects the clock to fire a bit later the next time around. (In this way, the cricket nervous system works very much like Huygens's pendulum clocks, with negative feedback producing the kinds of alterations that foster synchrony.) If we could quantify the chirp rhythm of many individual crickets in isolation, and describe how each one modifies its rhythm in response to the chirps of others, our mathematics should be able to predict their communal behavior over a wide range of conditions.

Tim built ingenious little soundproofed boxes, each to house one cricket. Every box had its own tiny microphone for piping the sound of its inhabitant's chirp to other crickets, and one tiny speaker for listening to incoming signals. This contrived experimental design allowed us to control the interaction strength—we could amplify the chirps or play them as softly as we wished—and we could even control which crickets could hear which, by hooking the boxes together in various patterns of connectivity.

As Duncan pondered the possibilities, he began to think about connectivity more generally. Out in the field, it wasn't obvious which crickets were listening to which. If anything, they seemed to be scattered in the trees, forming no discernible pattern. A male might attend only to his nearest competitors. Or maybe he listened to all the others. It wasn't even clear if the connectivity mattered; maybe they'd synchronize in any case.

One day in January 1996, Duncan walked into my office with an offbeat idea—yet another change of direction for his research. While daydreaming about connectivity, he suddenly remembered something his father once said, about how we're all just six handshakes from the president of the United States. He wondered if that were really true, and if so, what it meant about how the world was connected.

Sure, I said, I've heard of six degrees of separation. It was a math problem in disguise, waiting to be formulated.

But that's not all, Duncan went on. Six degrees of separation is related to what we're trying to figure out about the crickets. Suppose a network of biological oscillators is connected in such a way that everyone is a few handshakes apart. Does that affect how the group will synchronize? Will it synchronize very fast and very easily, because everyone is so tightly linked? Will the system still show a phase transition as the coupling strength is increased, like the Kuramoto model does?

Nobody knows about that, I told him, we never study those kinds of networks. And that was his point. Oscillator theorists had always assumed that their networks were perfectly regular, as orderly as the atoms in a crystal. Winfree and Peskin and Kuramoto had all built models with maximum connectivity, with every oscillator coupled to every other. No network could be more densely connected; no architecture could be simpler. In later extensions of those models, mathematicians had laid the oscillators end to end in a long chain, or positioned them symmetrically at the corners of a square grid or a three-dimensional lattice. Regular geometries like these were justifiable for problems coming from physics and engineering; in arrays of Josephson junctions, for example, the superconducting oscillators are deliberately fabricated in neat rows and columns. Even in a continuous medium like a beaker full of the Belousov-Zhabotinsky reaction, the connectivity is still regular: The chemicals diffuse primarily to their nearest neighbors.

On the other hand, for the tangle of neurons in the brain, where cells connect extensively to others nearby but also send long-range fibers halfway across the cortex, grids and lattices were clearly inappropriate. A better model of the geometry would incorporate a looser, more relaxed kind of structure, a hodgepodge of order and randomness, with clustered local connections and haphazard global ones. Maybe the same was true for crickets. Maybe there was a whole new class of oscillator networks waiting to be analyzed.

I was skeptical. Coupled oscillators on regular grids were already formidable; these new hybrid networks would be hopeless. But I liked Duncan's initiative and didn't want to stifle him.

As we began to bat his idea around, I started to appreciate his larger point. The same issues were bound to arise for other kinds of dynamical systems, not just coupled oscillators. Whenever nonlinear elements are hooked together in gigantic webs, the wiring diagram has to matter. It's a basic principle: Structure always affects function. The structure of social networks affects the spread of information and disease; the structure of the power grid affects the stability of power transmission. The same must be true for species in an ecosystem, companies in the global marketplace, cascades of enzyme reactions in living cells. The layout of the web must profoundly shape its dynamics.

Yet theorists had typically skirted the issue of connectivity. When they weren't assuming unrealistically regular arrangements, they lurched to the opposite extreme, modeling the connectivity as totally random. For example, in 1969, the theoretical biologist Stuart Kauffman proposed an idealized model of gene networks in which each gene was regulated by the products of two others, selected at random from the rest of the genome: not because he believed it was really like that, but because in 1969 nothing was known about how gene networks were actually wired. The random assumption is a way of throwing up one's hands, a null hypothesis in the absence of any information. Mathematical epidemiologists often resorted to the same approximation; they'd assume that infected people mixed at random with susceptible ones, even though for many kinds of diseases (especially sexually transmitted ones), the network of contacts couldn't possibly be random. Like regular networks, random ones are seductive idealizations. Theorists find them beguiling, not because of their verisimilitude, but because they're the easiest ones to analyze.

By 1996, the twin fictions of regular and random networks were starting to look less plausible every day. Anyone surfing the Web could sense it was both a pattern and a maze, where Web pages link mainly to others on the same topic but occasionally veer off onto idiosyncratic byways. AIDS and Ebola demonstrated that infectious diseases spread mainly within tight-knit communities, but also hitch rides on airplanes. So it was fitting that Duncan was now proposing a trek into uncharted territory, to the realm between order and randomness.

We began by trying to visualize what a network "in the middle" might look like. The simplest approach is to take a regular network and smoothly transform it into a random one (somewhat like the Hollywood special effect of morphing one face into another, as in Michael Jackson's video *Black or White*). For instance, halfway through the transformation, we would pick half of the original connections in a network, delete them, and replace them with an equal number of links strewn between random pairs of nodes. The resulting network would still have the same number of links as the original, but it would now be half random, half regular. Or instead of rewiring half the links, we could choose any other fraction between 0 and 1. By dialing in any desired amount of rewiring, we could gradually tune the network from 0 (the original, pristine network with no rewiring) to 1 (a completely rewired, random mess). In between, the network would be an adjustable blend of the two.

As a concrete example, consider 6 billion nodes arranged in a circle. These nodes represent computers, neurons, people—whatever the components of the network happen to be. Imagine that each node is connected to exactly 1,000 neighbors: 500 nodes on its left, 500 nodes on its right. This is an extremely ordered network, a beautifully symmetrical ring lattice. At this stage the tuner is set to 0, the regular end of the spectrum. Now slowly turn the knob up from 0 to begin the morph. A few links break free from their moorings and redistribute themselves haphazardly. As the metamorphosis continues, more and more links change to random connections, eroding the symmetry of the original ring lattice, while leaving part of its structure intact.

We introduced two statistics to quantify the network's evolving architecture. One of them, the "average path length," formalizes the intuitive idea of degrees of separation. To calculate it, take any pair of nodes and count the number of links in the shortest chain between them; then repeat for all other pairs of nodes, and average the resulting chain lengths.

For the pristine ring this calculation is easy, especially if you picture the network as a society, where the nodes represent people and the links represent friendships. This imaginary world ("RingWorld") is a bit like our own in one way—it too has 6 billion people—but otherwise it's completely alien. Its inhabitants are condemned to live in outlandishly regimented conditions, with everybody standing shoulder to shoulder in an astronomically huge circle. Let's say then that each person is forced to befriend the 500 people on his left, and the 500 people on his right, but no one else. A world like this wouldn't have six degrees of separation; it would have a whopping 3 million.

To see why, consider the path length between you and the most distant person on the ring, diametrically opposite you on the far side. To reach him by the shortest chain, you'd send a signal out to your 500th friend (the closest one to the target). From there, the fastest path would again hop over all the intervening people, out to that person's 500th friend, and so on, leapfrogging around the ring in chunks of 500 people. The entire journey requires 3 billion divided by 500 steps, which is 6 million steps. But that was for the most distant possible target. For the closest target—a person standing next to you—only one step is required. So on average, the distance between you and a typical person is about 3 million handshakes, 3 million degrees of separation.

At the other end of the spectrum, when the morph is over and the network has become totally random, the calculation is equally straightforward. Now, remarkably, everyone is only four steps from everyone else. The explanation has to do with exponential growth. In a random world, if you know 1,000 people (on average), and each of them knows 1,000 people, that means there are 1 million (= 1,000 × 1,000) people within two steps of you, 1 billion within three, and 1 trillion—much more than the world's population—within four.

It's tempting to extrapolate the same argument to our own world, to explain how we could all be six degrees of separation apart, but here the argument fails. It overlooks the fact that real friendship circles overlap—that many of your friend's friends are your friends too, and are therefore being counted twice.

For a hypothetical network that's both sparse and totally random, however, the calculation is valid, because the overlap of friendship circles is negligible in this case. When you pick 1,000 people at random from a vast pool of 6 billion, and so do all your friends, the chances of any overlap are only 1 in 6 million, as it turns out. So there's essentially no double counting. Such a world would be bizarre, of course—you'd be just as likely to know a peasant in the Himalayas, the Prince of Wales, or the person next door. Your friends would be scattered across every continent and class, every race and religion. In a world with no overlap, there'd be no social structure, no families, no communities.

This argument highlights the importance of understanding the concept of overlap more generally. The average amount of overlap in a network is quantified by a second statistic, the "clustering," defined as the probability that two nodes linked to a common node will also be linked to each other (or in human terms, the probability that friends of a friend are also friends of each other). In the two extreme models discussed above, the clustering can be shown to vary from a towering high of 0.75 for the pristine ring to a subatomic low of 1 in 6 million for the random net.

To arrive at the number 0.75, for example, you need to recognize that you have virtually all the same friends as a person standing right beside you on the ring (998 out of 1,000, to be exact), so your overlap with that *closest* person is essentially equal to 1. On the other hand, with your most *distant* friend, 500 steps away on the ring, you share only about half the same friends (they are the 499 people who happen to be sandwiched between you both on the ring), so your overlap with that most distant person is 499/1,000, which is essentially ½. For all your other friends lying between the closest and farthest one, the overlap changes smoothly from ½ to 1, yielding an average of ¾, which is the 0.75

value for the clustering quoted above. Next, we can apply similar but somewhat easier reasoning to calculate that if the connectivity were random, the clustering would equal 1 in 6 million; have fun working it out if you're curious. But please don't get lost in these details. The crucial point here is that, just like the average path length, the clustering plunges almost a millionfold as we morph the network from one end of the spectrum to the other.

Although the two statistics drop by a similar factor, they track very different aspects of the network's architecture. Average path length reflects the global structure; it depends on the way the entire network is connected, and cannot be inferred from any local measurement. Clustering reflects the local structure; it depends only on the interconnectedness of a typical neighborhood, the inbreeding among nodes tied to a common center. Roughly speaking, path length measures how big the network is. Clustering measures how incestuous it is.

So far we have concentrated on the traditional ends of the network spectrum. But we're still in the dark about what happens in the middle. The endpoints alone tell us that the morph somehow shrinks the ring enormously and destroys its clusters. What remains unknown is whether the transition is gradual or abrupt. Neither Duncan nor I could see how to solve that problem by pure mathematics, so we used a computer to simulate the morph on networks of large but manageable size, starting from pristine rings with 1,000 nodes and 10 links per node. To chart the structural changes in the middle ground, we graphed both the average path length and the clustering as functions of the proportion of links that were randomly rewired.

What we found amazed us. The slightest bit of randomness contracted the network tremendously. The average path length plummeted at first—with only 1 percent rewiring (meaning that only 1 out of every 100 links was randomized), the graph dropped by 85 percent from its original level. Further rewiring had only a minimal effect; the curve leveled off onto a low-lying plateau, indicating that the network had already gotten about as small as it could possibly get, as if it were completely random. Meanwhile, the clustering barely budged. With 1 percent rewiring, the clustering dropped by only 3 percent. Connections were being yanked out of well-ordered neighborhoods, yet the clustering hardly noticed. Only much later in the morph, long after the crash in path length, did clustering begin to drop significantly.

These results have an intuitive explanation. At the beginning of the morph, the first few random links act as shortcuts—bridges between parts of the network that would otherwise be remote. Their disproportionate impact comes from a powerful nonlinear effect: Not only do they pull two nodes together; they pull entire worlds together. For example, I like to play chess on-line at the Internet Chess Club, where I've become friendly with Emilo, the editor of a magazine in Holland. Thanks to that shortcut, I'm much closer to him of course, but also to thousands of other people in Holland—all his friends, and friends of those friends—than I was before. And though my friends don't realize it, all of them are now closer to him and his friends, through the single shortcut we forged. That one bridge does a lot of work.

In our simulations, the first few shortcuts drastically reduced the size of the world, but had far less effect on the clustering. The implication is that the transition to a small world is essentially undetectable at a local level. If you were living through the morph, nothing about your immediate neighborhood would tell you that the world had become small. You'd still have the same number of friends, with no sense about whether they connected you to a wider circle. Someone in a world like this might feel insulated from the threat of a disease like AIDS—rationalizing, for example, that none of his sexual partners were in high-risk groups—though in reality he might be just a step or two removed.

The most important result of the simulations was that over a broad intermediate range of rewiring, the model networks were very clustered and very small at the same time. That peculiar combination was new to mathematics. In traditional networks, size and clustering go hand in hand. Random networks are small and poorly clustered; regular ones are big and highly clustered. The rewired networks managed to be both small and highly clustered simultaneously.

We dubbed networks with this pair of seemingly contradictory properties "small-world networks," in homage to the same duality that seems so paradoxical about human connectedness: We move in tight circles yet we are all bound together by remarkably short chains. The question now was whether nature makes use of this strange form of network architecture, and if so, to what end.

Our simulations suggested that small-world connectivity should be widespread in real networks, since even a tiny fraction of shortcuts would suffice. To test that prediction, we needed empirical examples. They weren't easy to find. Any candidate had to be fully characterized, its wiring diagram known down to the last detail, every node and link documented, or we couldn't calculate the clustering and average path length.

Then I remembered that Koeunyi Bae, a student in my chaos course the year before, had done a project about the Western States power grid, a collection of about 5,000 electric power plants tied together by high-voltage transmission lines across the states west of the Rocky Mountains and into the western provinces of Canada. Koeunyi and her adviser Jim Thorp provided the data to Duncan. It contained a great deal of detailed information that an engineer would find crucial—the voltage capacity of the transmission lines, the classification of the nodes as transformers, substations, or generators—but we ignored everything except the connectivity. The grid became an abstract pattern of dots connected by lines. To check whether it was a small-world network, we compared its clustering and average path length to the corresponding values for a random network with the same number of nodes and links. As predicted, the real network was almost as small as a random one, but much more highly clustered. Specifically, the path length was only 1.5 times larger than random, whereas the clustering was 16 times larger.

Turning from technological networks to biological ones, we next looked at the nervous system of a tiny worm called C. elegans. More is known about this humble creature—a transparent, soil-dwelling nematode only about a millimeter long—than about any other animal, even including the geneticist's fruit fly and the oncologist's mouse. Every one of the worm's 959 cells has been mapped at every stage of its development, from conception to death. Its entire genome was sequenced as long ago as 1998. Abstruse as this organism may seem, its study has illuminated several fundamental cellular processes, from cell death to cell signaling to the guidance of nerve axons, all of which were first discovered by worm biologists and later shown to have significance for humans. And that's precisely why so much attention has been lavished on *C. elegans:* It is perhaps the simplest organism that shares many of the biological processes essential to human life.

For our purposes, the attraction of *C. elegans* was that its nervous system had been completely mapped, a feat not yet achieved for any other organism. In fact, the wiring diagram for its 302 neurons was readily available on a floppy diskette. As with the power grid, we neglected the details that a specialist would find most meaningful. We treated the neurons as identical (even though biologists distinguish among 118 different classes), and regarded two neurons as connected if they're linked by either a synapse (a chemical connection) or a gap junction (an electrical connection).

The resulting abstract network again turned out to be a small world. Its average path length was a mere 18 percent larger than that of a corresponding random net, whereas its clustering was six times larger. What this meant was unclear. It could be that the short path length facilitates rapid communication throughout the creature's body, while the high clustering probably reflects the presence of feedback loops and modular structure in its nervous system.

Two radically different networks, the power grid and the nervous system: one created by mankind, the other by evolution. One is among the largest machines ever built, a sprawling web of synchronized generators linked by hundreds of thousands of miles of cable. The other is a microscopic filigree, the product of millions of years of natural selection, a lacework snuggled in the body of a worm. And yet despite all their differences, their architecture is strikingly similar. Both networks are almost as small as they could possibly be. Both are highly structured and definitely not random. Admittedly, our approximations clouded the interpretation of these findings—the small-world architecture of both networks might be irrelevant to their function, and therefore meaningless. Time would tell. But for now the coincidence was tantalizing.

. . .

Social networks also seemed likely to be small worlds, though we were unaware of any supporting evidence beyond the anecdotal. In particular we wondered if the notion of "six degrees of separation" was based on hard, verifiable data. Perhaps it was nothing more than an urban legend. (John Guare himself couldn't remember—he thought it might have come from Guglielmo Marconi, inventor of the wireless telegraph, back in the years when he was connecting the planet with telegraph stations.) Without real data, we couldn't be sure if our theory was as broadly applicable as we suspected it might be. Did it apply to networks of human beings, as well as to power grids and nervous systems?

Our first lead came from a conversation with Joel Cohen, a mathematical biologist at Rockefeller University who'd used network theory to model the structure of ecological food webs. When I mentioned that we were trying to educate ourselves about the empirical basis for six degrees of separation, he said, "You mean the small-world problem" and directed us to the classic work of Stanley Milgram.

In 1967, Milgram, a social psychologist at Harvard, devised an experiment to test whether American society was more like an array of disconnected islands or one giant, interlocking web. The experiment was intended to trace a line of acquaintances between two randomly chosen people in the United States. Milgram gave a folder (an impressive-looking booklet, somewhat like a passport with the Harvard seal embossed on it) to a person at the start of the chain, with instructions to send it toward a designated target person, but with a caveat: "If you do not know the target person on a personal basis, do not attempt to contact him directly. Instead, mail this folder . . . to a personal acquaintance who is more likely than you to know the target person . . . it must be someone you know on a first-name basis." In this way the folder would march its way across the country from acquaintance to acquaintance, gradually zeroing in on the target. To initiate the chains, Milgram solicited volunteers from the Midwest, for reasons he later explained: "As a crude beginning, we thought it best to draw our starting people from some distant city such as Wichita, Kansas, or Omaha,

Nebraska (from Cambridge, these cities seem vaguely 'out there,' on the Great Plains or somewhere)." The Nebraska study involved 160 starting people, all of whom were trying to reach the same target, a stockbroker who lived in Sharon, Massachusetts, and worked in Boston. At the time, Milgram wasn't sure any of the chains would get through, or how many steps they might require. "When I asked an intelligent friend of mine how many steps he thought it would take, he estimated that it would require 100 intermediate persons or more to move from Nebraska to Sharon," Milgram wrote.

The result: After passing through only 2–10 intermediate acquaintances, 44 folders successfully reached the target. The median number of intermediaries was 5, corresponding to 6 links in the chain—the number now enshrined in popular culture as six degrees of separation. (The other chains weren't completed, because some people couldn't be bothered to cooperate and pass the folder along.)

Intriguing as these results are, they remain inconclusive. The chains might not have been the shortest ones possible, so the true average path length can't be estimated. It could even be infinite: There could be pairs of people in the United States who live in unbridgeable social universes, with no chains between them. And without more information about the network's local connectivity, it was impossible for us to calculate its clustering. To answer these more detailed questions, we still needed to find a social network that was fully characterized, with every node and link documented beyond dispute.

Mathematicians themselves had jokingly begun such an enumeration, centering their universe around Paul Erdős, a Hungarian savant who was utterly incompetent at all aspects of everyday life—he couldn't (or wouldn't) even butter his own toast. Yet Erdős was one of the most prolific and inventive mathematical minds of the twentieth century. High on amphetamines, wandering around the world with nothing but his beaten-up old suitcase, he'd show up at your doorstep and announce, "My brain is open," meaning he was ready to work on an unsolved math problem with you.

Erdős collaborated with so many people that it became a popular game among mathematicians to compute your "Erdős number." If you're one of the honored few to have co-authored a paper with him (there are 507 such people), you have an Erdős number of 1. If you have never written a paper with Erdős himself, but you have written one with someone who has, then you have an Erdős number of 2. The joke in mathematical circles was that anyone who's anyone will have an Erdős number of 2 or less. There's a Web site that lists all the people lucky enough to be 1s and 2s, but no list of 3s is available. It would be enormous. (I'm among them.) Unfortunately, without the full list, we couldn't calculate average path length or clustering for this social network either. Human networks were proving to be frustratingly elusive.

Whenever we described our work to laypeople, they invariably brought up the Kevin Bacon game. We'd always laughed that off, but now we began to see an opportunity here, a way out of our quandary. The network of movie actors could be a surrogate for a social network. Instead of people connected by friendships, the net would consist of actors connected by movies. Two actors who've appeared in the same film are considered to be one step apart; if they've never been in a film together but have a common costar, they're two steps apart, and so on. Though a bit whimsical, this network had the advantage of being comprehensive. The Internet Movie Database includes the cast of virtually every feature film ever made. On the other hand, its size would also cause a problem: As of April 1997, the network contained nearly a quarter of a million actors, so the calculation would be gigantic. Even Cornell's supercomputer, one of the largest in the world, was going to have trouble holding all the data in memory.

Fortunately, Brett Tjaden (aka The Oracle of Bacon), a computer scientist at the University of Virginia, had already spent several weeks computing the shortest chain of movies between any pair of actors. Along the way he found that the network has an interesting global structure. It's dominated by one enormous, connected piece (known as the "giant component") with 90 percent of all actors in it, including Kevin Bacon and every other actor you've ever heard of. But it also contains a smattering of tiny islands, pockets of obscure actors cut off from the rest of the acting universe (for example, people who appeared in one movie that they made in film school with all their friends, none of whom ever acted in another movie again).

Using Tjaden's data, Duncan computed that any two randomly chosen actors in the giant component are separated, on average, by just 3.65 movies: an impressively small number, considering that the actors come from every country, genre, and era, from silent films to the present. If the network had been completely random, the corresponding number would have been smaller, but not much: 2.99. The clustering, on the other hand, turned out to be extraordinarily large: 0.79, about 3,000 times larger than the value for a random net.

So we were seeing the same duality again: short chains and high clustering, the signature of a small-world network. For whatever reason, maybe luck or maybe a hint of something deeper, we were now three for three. Each of the networks we had looked at (and they were not handpicked) had turned out to be small worlds. That similarity was especially striking in light of the networks' disparate sizes and scientific origins. It was starting to seem like small-world architecture might be remarkably pervasive.

Incidentally, the analysis also toppled Kevin Bacon from his pedestal. He ranked number 669 on the list of best-connected actors, as measured by his average separation from everyone else in the giant component. By this measure, the center of the Hollywood universe is Rod Steiger. Unexpectedly, number 2 and number 3 are Christopher Lee and Donald Pleasence, best known for their work in cheesy horror films.

Having demonstrated that small-world networks not only exist, but that they might even be ubiquitous, we still needed to address Duncan's original question: Would oscillators coupled in a small-world fashion synchronize more or less readily than they would in a traditional, regular network? That issue could finally be addressed, at least theoretically, with the help of the morphing model developed earlier. Each node in the network would now represent a self-sustained oscillator—a singing cricket, a flashing firefly, a pacemaker neuron—and the links would reflect the pattern of interactions.

One of the simplest models of this sort had previously been studied by Kuramoto and his colleagues Hidetsugu Sakaguchi and Shigeru Shinomoto. They'd considered the same kinds of oscillators as in the original Kuramoto model—phase oscillators with distributed natural frequencies, mutually coupled by an attractive sine-wave interaction. (Think of a roomful of people trying to applaud in unison by speeding up or slowing down—depending on their timing relative to the collective clap—in an attempt to overcome their diverse clapping speeds, which run the gamut from stately to frenetic.) But unlike the original Kuramoto model, where the oscillators were coupled all-to-all, the Japanese physicists now assumed a ring of connectivity, with oscillators arranged in a circle, each coupled to a fixed number of neighbors on either side. (Picture a circular arena—a football stadium—where each fan listens exclusively to others sitting next to him.) Kuramoto and his colleagues found that a ring of dissimilar oscillators could not easily achieve widespread synchrony; it tended to fragment into many small groups of neighbors, all cycling at the same average speed within a group, but varying from group to group. Different sections of the stadium would now be clapping at different rates.

We wondered if rewiring the ring might enhance its ability to synchronize. As in earlier simulations, we morphed the ring lattice toward a random net by converting some of its original connections to random ones. (It was as if a few fans had cellular phones, piping in the applause from remote parts of the stadium that none of their section mates could perceive.) We found that a tiny percentage of such shortcuts—on the order of 1 or 2 percent in a ring of 1,000 oscillators—changed the overall dynamics dramatically. The system flipped spontaneously from parochial discord to global consensus. Now all the oscillators locked their rhythms to a single compromise frequency.

Though we couldn't see how to explain these results mathematically, an intuitive explanation suggested itself: The shortcuts were providing high-speed communication channels, enabling mutual influence to spread swiftly throughout the population. Of course, the same effect could have been achieved by connecting every oscillator directly to every other, but at a much greater cost in wiring. The small-world architecture apparently fostered global coordination more efficiently.

By the same token, perhaps small-world architecture would be advantageous in other settings where information needs to flow swiftly throughout an enormous complex system. The test case we studied next is a classic puzzle in computer science called the "density classification problem for one-dimensional binary automata." In plainer language, imagine a ring of 1,000 lightbulbs. Each bulb is on or off. In the next time step, each bulb looks at its three neighbors on either side, and using some sort of clever rule (to be determined), it decides whether to be on or off in the next round. The puzzle is to design a rule that will allow the network to solve a certain computational task, one that sounds ridiculously easy at first: to decide whether most of the bulbs were initially on or off. If more than half the bulbs were on, the repeated execution of the rule is supposed to drive the whole network to a final state with all bulbs on (and conversely, if most bulbs were off at the start, the final state is supposed to be all off).

The puzzle is trivial if there is a central processor, an eye in the sky that can inspect the whole system and count whether most bulbs were initially on or off. But remember, this system is decentralized. No one has global knowledge. The bulbs are myopic: They can see only three neighbors on either side, by assumption. And that's what makes the puzzle so challenging: How can the system, using a local rule, solve a problem that is fundamentally global in character?

This puzzle captures the essence of what's called collective computation. Think of a colony of ants building a nest. Individually, no ant knows what the colony is supposed to be doing, but together, they act like they have a mind. Or recall Adam Smith's concept of the invisible hand, where, if everyone makes a local calculation to act in his or her self-interest, the whole economy supposedly evolves to a state that's good for all. Here, in the density classification problem, similar (but much simpler) issues can be addressed in an idealized, well-controlled setting. The challenge is to devise a rule that will allow the network to decide whether most bulbs are initially on or off, for any initial configuration. The network is allowed to run for a time equal to twice its length. If there are 1,000 bulbs, the system is allowed to execute its local rule for 2,000 steps before it has to reach a verdict.

No one has yet found a rule that works every time. The world record is a rule that succeeds about 82 percent of the time—that is, it correctly classifies about 82 percent of all initial conditions as "more on" or "more off" within the allotted time. The first rule you might think to try—majority rule, where each bulb apes whatever the majority of its local neighborhood is doing—never works. The network locks up into a striped state, with blocks of contiguous bulbs that are on, interdigitated with blocks of bulbs that are off. That result is unacceptable, like a deadlocked jury. The net is supposed to converge to a unanimous verdict, with all bulbs either on or off.

Duncan and I guessed that a small-world network of bulbs might be able to solve the problem more efficiently than the original ring lattice. Converting a few of the links to random shortcuts might allow distant bulbs to communicate quickly, possibly preventing the hang-up in the striped state. We studied the performance of majority rule on ring networks with various amounts of random rewiring. As expected, when there was very little rewiring, majority rule continued to fail; the system was indistinguishable from a pristine ring, and again blundered its way into a deadlocked striped state. As we increased the amount of rewiring, the network's performance remained low for a while, but then jumped up abruptly at a certain threshold—at about the place where each bulb had one shortcut emanating from it, on average. In this regime, majority rule now began to perform brilliantly, correctly classifying about 88 percent of all initial configurations. In other words, a dumb rule (majority rule) running on a smart architecture (a small world) achieved performances that broke the world record.

The network spontaneously developed the ability to compute, once its wiring diagram was altered in a subtle way. The suggestion is that small-world architecture may be a powerful design for other problems of collective computation, one that confers surprising strength on even simpleminded local rules. As such, it's tempting to speculate that evolution might exploit this architecture in its design of biological nervous systems.

The importance of small-world connectivity is even clearer for processes involving contagion. Anything that can spread—infectious diseases, computer viruses, ideas, rumors—will spread much more easily and quickly in a small world. The less obvious point is how few shortcuts are needed to make the world small.

The awesome reach of shortcuts was tragically illustrated by the spread of AIDS through North America, believed to have been hastened by Patient Zero, a promiscuous French-Canadian flight attendant who traveled worldwide and frequented bathhouses in San Francisco, Los Angeles, Vancouver, Toronto, and New York. At least 40 of the first 248 men diagnosed with AIDS had sex either with him or with one of his previous partners.

Similarly, epidemiologists in the United Kingdom have noticed an alarmingly new pattern of spread in the latest outbreak of foot-and-mouth disease, a highly contagious virus that afflicts cows, pigs, sheep, and other cloven-hoofed animals, with devastating economic consequences for the livestock industry. During the last outbreak in 1967, the disease propagated mainly by airborne-particle diffusion (though it can also be carried by birds and animals, and even on shoes and clothing). Of the roughly 2,000 cases, more than 95 percent were localized within 100 kilometers of the source of the outbreak. In contrast, the current epidemic already extends over a 500-kilometer range within the United Kingdom. The difference is thought to be due to changes in agribusiness, especially the increased transport of livestock between distant dealerships and markets—the shortcut mechanism for this disease. The virus has already spread from England to Ireland, France, and Holland, and since the year 2000 alone, outbreaks have been reported in 34 countries. Although foot-and-mouth disease has not yet entered the United States (as of this writing), and hasn't struck here since 1929, there is no cause for complacency. As two commentators recently put it, "We are not just living in a 'global village'; we are living on a global farm."

The propagation of computer viruses and worms on the Internet also demonstrates the efficacy of small-world connectivity. Consider the Love Bug worm, which automatically forwarded itself to everyone on a victim's E-mail list. Given that the on-line community is probably clustered into tight, inward-looking circles of friends and associates, it's a bit surprising that the worm managed to infect so many of the world's computers in a matter of days; one might have expected it to circulate endlessly within a narrow community. Presumably a few long-range connections enabled it to leap from one social world to another.

On a happier note, shortcuts also have beneficial uses in our everyday lives. In the late 1960s, the sociologist Mark Granovetter asked hundreds of professionals and technical workers how they found their jobs. As he recalled during a radio interview,

> When I started interviewing people about how they found jobs, of course I found that they often found jobs through personal contacts, and I was interested in who these contacts were, and how the information was flowing, and why it was flowing, and I would often say to these people, "Was this a friend you got the information from?" and they kept correcting me and saying, "No, no, it was just an acquaintance." And I realized after a while, after people kept saying this to me, that there was suddenly something systematic here. And the fundamental idea is that your close friends are wonderful for all kinds of things—for giving you support, for helping you when you're sad, for doing favors that other people wouldn't do for you—but as sources of information they're not very good, because your close friends tend to know the same people you know. Whereas people who are just your acquaintances—who might not help you out if you were in desperate trouble—are still better sources of information because they know so many people you don't know. They're really your windows on the world, because they're linked up to different circles from your own.

Specifically, Granovetter found that of the 56 percent of people who found their jobs through personal contacts, only 17 percent saw that contact "often" (as they would have, had the contact been a good friend), whereas 55 percent saw their contact "occasionally" and 28 percent saw the contact "rarely." Granovetter invented a memorable phrase to describe the vital function of these relationships outside one's usual orbit. His now-famous paper is titled "The Strength of Weak Ties."

While Duncan and I were exploring small-world networks and their possible implications, another team was independently thinking along similar lines.

At the University of Notre Dame, László Barabási and his students Réka Albert and Hawoong Jeong were probing the anatomy of the World Wide Web, searching for regularities in this bewildering thicket of a billion pages connected by hyperlinks. What they uncovered has turned out to be yet another organizing principle for a broad class of natural and man-made networks.

Barabási is an energetic young physicist with a delightful Transylvanian accent and a flair for asking the right questions. Trained in statistical mechanics (the branch of physics that deals with enormous systems of atoms and other collections of particles), he brought a novel set of tools to a puzzle outside the purview of conventional physics. He and his team showed that the Web is not only a small world, but that it displays a peculiar pattern in its anatomy. Some pages are much more highly connected than others, with many more incoming or outgoing links than average. That much was not surprising: Any population is bound to contain some outliers at the far ends of the spectrum. But what was surprising was the shape of the distribution. It was not a familiar bell curve, like the distribution of human heights. It was more like the distribution of incomes, with a monstrously long tail extending to the right. (The implications of this peculiar structure are explored extensively in Barabási's recent book *Linked*.)

In the distributions studied in traditional statistics courses, the average value sets a characteristic scale, a typical size for the members of the population as a whole. For example, consider the distribution of human heights. Nearly every adult is between two and nine feet tall. You never meet someone an inch tall or a hundred feet tall. Human heights have a characteristic scale of around five feet, and certainly don't deviate from it by more than an order of magnitude (a factor of ten) on either side of the mean. In contrast, the income distribution spans many orders of magnitude, from yearly incomes close to zero, all the way up to the billions of dollars that Bill Gates makes on interest alone. A distribution like this is sometimes called "scale free," meaning that it is not dominated by any single, representative scale.

What Barabási and his collaborators discovered is that the distribution of links on the Web is similarly scale free, and for the same reason—the curve has an outrageously long and heavy tail. Specifically, the tail decays at a much slower rate than a normal bell curve. Instead of decaying exponentially fast, it tapers off according to a "power law" with an exponent of 2.2. In algebraic terms, the law says that for every tenfold decrease in the number of incoming links, the number of pages having that number of links will increase, on average, by a factor of 10 raised to the 2.2 power, which is roughly equal to 158. Or to put this the other way around, pages with 10 times more links will be 158 times less likely.

This arcane pattern holds across the entire Web, from a handful of giant hubs like CNN and Yahoo, each with thousands of incoming links, to the hundreds of billions of pages languishing in obscurity, with no incoming links at all. From a purely mathematical perspective, a power law signifies nothing in particular—it's just one of many possible kinds of algebraic relationship. But when a physicist sees a power law, his eyes light up. For power laws hint that a system may be organizing itself. They arise at phase transitions, when a system is poised at the brink, teetering between order and chaos. They arise in fractals, when an arbitrarily small piece of a complex shape is a microcosm of the whole. They arise in the statistics of natural hazards—avalanches and earthquakes, floods and forest fires—whose sizes fluctuate so erratically from one event to the next that the average cannot adequately stand in for the distribution as a whole. But despite 20 years of intense effort, the origin of power laws remains controversial.

For all these reasons, the discovery of a power law in the Web came as a shock and a provocation. The Web is an unregulated, unruly labyrinth where anyone can post a document and link it to any other page at will. There was no reason to expect any pattern at all. And yet the Web is apparently ordered in a subtle and mysterious way, following the same power-law pattern that keeps popping up everywhere.

Barabási and his team offered an intriguing explanation. In their eyes, the power law is a natural consequence of network growth. The Web is not static. New pages are born every day, links are added, rewired, and lost, and old pages die. Suppose, to a rough approximation, that all these processes can be ignored except for the addition of new pages, and that new pages link at random to existing ones, but with a preference for pages that happen to be popular. Then richly connected nodes get richer, and a mathematical analysis shows that a power law emerges automatically with an exponent of 3, not far from the observed value of 2.2. More refined models have since narrowed that gap.

In the past five years, the new ideas of small-world and scale-free networks have triggered an explosion of empirical studies dissecting the structure of complex networks. In case after disparate case, when the flesh is peeled back, the same skeletal structure appears from within. The Internet backbone and the primate brain—both small worlds. So are the food webs of species preying on each other, the meshwork of metabolic reactions in the cell, the interlocking boards of directors of the Fortune 1,000 companies, even the structure of the English language itself. Most of these networks, though not all, are scale free as well (that is, more like the income distribution and less like the height distribution).

At an anatomical level—the level of pure, abstract connectivity—we seem to have stumbled upon a universal pattern of complexity. Disparate networks show the same three tendencies: short chains, high clustering, and scale-free link distributions. The coincidences are eerie, and baffling to interpret.

For example, to construct a network for the English language, the physicists Ramon Ferrer i Cancho and Ricard Solé considered two words to be linked if they ever appear close together (either next to each other or one word apart) in sentences in the British National Corpus, a 100-million-word collection of samples of written and spoken language from a wide range of sources, designed to represent a cross section of current British English. Cancho and Solé found that you can hop from any word to any other in this way, in just 2.67 steps, on average. It seems at first like almost anything can happen (because reasonable English sentences are infinitely variegated), yet the linguistic network turned out to be highly organized and far from random, with a clustering of word associations more than 4,000 times greater than that of an equivalent random network. The wiring diagram of word associations is scale free with two distinct regimes: common words (those with more than 1,000 links) obey a power law with an exponent of 2.7, while for uncommon words the exponent equals 1.5.

In cases like this, it's unclear whether the patterns are genuinely significant or mere numerology. Admittedly, in all the excitement swirling around the subject of complex networks, there's been a tendency to make inflated claims. A physicist friend of mine ribbed me with his own mock discovery—a small-world pattern of icing on a piece of apple strudel tastes better *and* has fewer calories.

The challenge now is to decode the underlying meaning of small-world and scale-free architecture, if there is any. In one recent attempt, Solé has observed that electronic circuits tend to be wired in a small-world fashion, and he thinks he knows why. Whether he was analyzing the latest digital microchips or the clunky circuits found in old televisions, he found that all the components were just a few electrical steps from one another, yet they were much more clustered than they would have been in an equivalent random circuit, thanks to the modular design favored by engineering practice. Solé speculates that this kind of layout may have emerged by natural selection, as alternative designs competed for survival over time. In other words, engineers may have unknowingly built the circuits according to small-world principles, by trying to strike the best compromise between low cost and high reliability.

Barabási and his team pointed out that scale-free networks also embody a compromise bearing the stamp of natural selection: They are inherently resistant to random failures, yet vulnerable to deliberate attack against their hubs. Given that mutations occur at random, natural selection favors designs that can tolerate haphazard insults. By their very geometry, scale-free networks are robust with respect to random failures, because the vast majority of nodes have few links and are therefore expendable. Unfortunately, this evolutionary design has a downside. When hubs are selectively targeted (something that random mutation could never do), the integrity of the network degrades rapidly—the size of the giant component collapses and the average path length swells, as nodes become isolated, cast adrift on their own little islands.

Evidence for this predicted mix of robustness and fragility is manifested in the resilience of living cells. In a study of the network of protein interactions in yeast, Barabási's group found that the most highly connected proteins are indeed the most important ones for the cell's survival. They reached this conclusion by cleverly combining information from two different databases. First they looked at the connectivity data, where two proteins are regarded as linked if one is known to bind to the other. This interaction network follows a highly inhomogeneous, scale-free architecture, with a few kingpin proteins mediating the interactions among many more poorly connected peons. Then Barabási's team correlated the connectivity data with the results of systematic mutation experiments, in which biologists had previously deleted certain proteins to see if their removal would be lethal to the cell. They found that deletion of any of the peons (the 93 percent of all proteins having fewer than 5 links) proved fatal only 21 percent of the time. In other words, the cell is buffered against the loss of most of its individual proteins, just as a scale-free network is buffered against the random failures of most of its nodes. In contrast, the deletion of any of the kingpins (the top 1 percent of all proteins, each with 15 or more connections) proved deadly 62 percent of the time.

Soon after Duncan and I published our small-world paper in *Nature,* we were bombarded by the mass media, from the *New York Times* and *CBS News* to the Hungarian daily *Magyar Hírlap*. People from all walks of life began contacting us with their own thoughts and speculations. An article in *Business Week* suggested that small-world ideas could be used to redesign organizations, by adding a few shortcuts to improve the lines of communication between different levels in the hierarchy. Someone from Senator Paul Wellstone's office called, hoping to brainstorm about the best way to spread the word about the liberal senator from Minnesota, who was then entertaining a possible run for the presidency in 2000. The most memorable call was the one from the FBI forensic scientist who left a cryptic message on my machine, requesting that I phone back as soon as possible. With some apprehension I dialed the number. "Hair and fiber," said the voice at the other end. His question had to do with the secondary transfer of fibers. If a fiber found on the victim matches the sweatshirt worn by the suspect, the prosecutor will introduce that coincidence as evidence. Naturally the defense attorney will argue that thousands of similar sweatshirts were sold last year; maybe the victim picked up a stray fiber left behind by someone else previously sitting on the same seat on the bus. The question was whether, given the probability of such secondary transfers, the number of sweatshirts manufactured, the connectivity of American social networks, and any other relevant data, one could calculate the likelihood that the fiber did in fact come from the suspect.

I wasn't able to offer anyone much help.

In striving to understand the origins of spontaneous order, this infant theory of complex networks is another step on the long journey that began with Christiaan Huygens and his sympathetic pendulum clocks. After centuries of thinking about purely rhythmic entities—oscillators—coupled together two at a time, then all-to-all, then in regular networks in space, mathematicians and scientists have only just begun to consider more complex dynamics like chaos and excitability, and more complex architectures like small worlds and scale-free networks.

At this early stage, our models are pale imitations of reality. We pretend that networks are built from featureless, static, identical nodes, connected by links with no directionality and no diversity in their strength or character. Much still remains to be learned about pure connectivity, but it's also getting to be time to move on, to incorporate nonlinear dynamics into the networks, to look beyond minimalist wiring diagrams. The nodes in our models need to become oscillators, or neurons, or power plants. The links need to be diverse and dynamic themselves. We still know almost nothing about the laws governing the interactions between genes, or proteins, or people.

· *Ten* ·

THE HUMAN SIDE OF SYNC

O N A QUIET AFTERNOON IN THE SPRING of 1994, I was sitting in my office at MIT, immersed in a calculation, when a ringing phone dragged me back from the depths. "This is Jean calling from Alan Alda's office. Will you hold for a call from Mr. Alda?"

A few seconds later I heard that familiar voice. "Hello, this is Alan Alda. I don't know if you know me, I'm an actor."

"Yes?" I was dumbfounded.

"I just read your *Scientific American* article about synchronization, and I'd like to come talk to you about it."

He said he'd always been fascinated by fads, and he was wondering if they could be explained as a kind of human behavioral sync. It sounded pretty speculative to me, but I was intrigued. We arranged a visit, and I gave him directions to my office—enter at the Dome, walk down the Infinite Corridor, turn right at the Norbert Wiener poster and go to Building 2.

When he arrived, he launched into his idea, even before sitting down. He mentioned hula hoops and pet rocks, fads that seemingly came out of nowhere and spread infectiously. Within weeks, millions of people were swiveling their hips or doting on minerals. Just as abruptly, the crazes ended. How does this process work? And why do some ideas catch on while others flop? Is it just a matter of luck, or mass hysteria, or could there be an underlying logic to fads? If so, he felt we should try to understand that logic, because the same kind of social contagion that drives fads could be put to more serious uses. For example, a million children die each year of dehydration, even in villages where rehydration remedies are available; what if rehydration became "fashionable" among those children's mothers? When public health officials tried to promote the use of condoms in the Philippines, or to encourage girls in Africa to stay in school, they used popular songs and comic books to deliver the message, hoping to start an epidemic of social change. Although some real successes were achieved this way, they tended to be temporary. Perhaps a deeper understanding of fads would have helped create more lasting ones.

He had researched this topic extensively: read all the classical sociologists of crowd behavior and mob psychology, the marketing experts and advertising gurus, even the evolutionary biologist Richard Dawkins with his proposal that "memes" are the psychological equivalent of genes, contagious ideas competing for survival, with the winners proliferating through a cultural version of natural selection. Insightful as these suggestions were, Alan felt that no one had quite gotten to the bottom of the problem, and that fads were as perplexing as ever. What was missing was a detailed, testable theory of their dynamics. So when he read about coupled oscillators and the mathematical theory behind them, he began to wonder: Could the sudden emergence of a fad be analogous to the way that fireflies suddenly start blinking in unison?

His suggestion seemed plausible but difficult to formulate mathematically. The existing theory of synchronization was largely confined to *rhythmic* sync, where all the individuals are oscillators, always repeating the same cycle, predictable as pendulums. Human behavior could not be pinned down so easily. Plus, the only tractable connectivities were global, all-to-all networks, hardly relevant to the social networks through which fads propagate. And, most frustrating of all, the rules governing human interactions—the counterpart of coupling between oscillators—were unknown and possibly unknowable. I was disappointed to say it, but I couldn't see how to help him.

Still, we chatted for three more hours. The discussion ranged from evolution and psychology to chaos theory and quantum mechanics. When the conversation wound down, I offered to dazzle him with lunch at the MIT cafeteria in Walker Memorial. We picked up some food, found an empty table, and kept talking about science. A few students stared at us and whispered, and one of my colleagues came over to gawk, under the pretense of needing to ask me something. Eventually it became impossible to ignore a young student pacing back and forth, hovering at the far end of the table. Finally he approached and waited till we looked up.

"Um, excuse me."

"Yes?"

"Sorry, I just have to ask: Aren't you . . . Professor Strogatz?"

"Yes?"

"Oh, um, I just wanted to say I read your book about chaos and I really liked it."

Then he walked away. Alan and I looked at each other and burst out laughing.

Only at MIT . . .

Alan's question about fads underscored how little we know about the human side of sync. In the past, coupled oscillator theorists had shied away from questions of psychology and group behavior. Yet the signs of human sync are inescapable: the herd mentality of stock traders and the resultant booms and crashes in the market; the brutal stupidity of mobs; the political and business oversights caused by "group think"; and even such harmless curiosities as that awkward moment at a cocktail party when everyone simultaneously falls silent. These are all instances of sync at the level of the group. The psychological dimensions of sync also show up at the level of the individual: What is it about music that stirs us so? Or the spectacle of sync in nature, the graceful movements of flocks of birds and schools of fish? What is it about dancing together that gives us such pleasure? Why do we delight in coincidences?

In particular, at the time Alan brought up fads, not much was known about the mathematics of human group behavior. Apart from some pioneering work in the 1950s by Anatol Rapoport, and later efforts by mathematical sociologists and economists like Thomas Schelling—the discoverer of the "tipping point"—the field was hampered by a lack of empirical studies and mathematical tools, and by the embryonic state of computer simulation. In the past few years, however, the subject has undergone a renaissance. Sociologists are borrowing the techniques of network theory to analyze simple models of riots, fads, and the diffusion of innovations. Physicists have recently investigated how Eastern European concert audiences switch from disorganized clapping to thunderous, synchronized applause. Complexity theorists are developing new ideas about traffic flow, explaining why congestion can persist for hours—even in the absence of accidents or other apparent causes—or how a population of selfish drivers can inadvertently settle into a cooperative pattern of flow, where all vehicles move in tandem like a weird, congealed mass.

The findings of these studies are typically counterintuitive. Unanticipated forms of collective behavior emerge that are not obvious from the properties of the individuals themselves. All the models are extremely simplified, of course, but that's the point. If even their idealized behavior can surprise us, we may find clues about what to expect in the real thing.

The recent work on fads builds on a classic model developed by the sociologist Mark Granovetter in the 1970s. He illustrated his results with a story about a hypothetical mob involving 100 people, possibly on the brink of rioting. Granovetter assumed that each person's decision whether to riot or not is dependent on what everyone else is doing. Instigators will begin rioting even if no one else is. Other people need to see a critical number of others causing mayhem before they'll join in. That critical number—the person's threshold—is assumed to be distributed across the population according to some probability distribution.

Granovetter's most famous example concerns the case of a mob with a uniform distribution of thresholds ranging from 0 to 99. In other words, one person has threshold 0, another has threshold 1, and so on. It's easy to predict what will happen in a crowd like this. The person with threshold 0 is ready to begin rioting even if no one else is. He instigates the riot. Then the person with threshold 1 becomes activated, since he sees one person (the instigator) breaking windows. Now that two people are rioting, the person with a threshold of 2 joins in. Like the burning of a fuse, or the toppling of a row of dominoes, the riot recruits more and more people until everyone is involved. That much is obvious, but here's the twist. Suppose, said Granovetter, that we alter the initial composition of the crowd in the slightest way. Suppose the person with threshold 1 is replaced by someone with threshold 2. Now when the instigator starts looting, no one else joins him, since everyone's threshold is greater than 1. In other words, no riot.

The surprise here is that the two hypothetical situations are almost indistinguishable, at least by the usual sociological measures. The average makeup of the crowd has changed in the smallest way possible, and the overall distributions of thresholds are almost identical. Yet the outcomes are as divergent as they could be: an all-out riot in one case, a lone maniac on a rampage in the other. An onlooker might describe the first crowd as a bunch of thugs and the second as a peaceful demonstration marred by one lunatic, when in fact the two crowds are near replicas of each other. The lesson is that the collective dynamics of a crowd can be exquisitely sensitive to its composition, which may be one reason why mobs are so unpredictable.

Among the many simplifications in Granovetter's model, perhaps the most serious is that everyone is assumed to have perfect knowledge of everyone else. This approximation is the sociological analog of the all-to-all coupling we encountered in the simplest oscillator models, where every firefly can see every other. Duncan Watts (who has now gone on to become a professor of sociology at Columbia University) has recently worked out the mathematics for the more realistic case where everyone is influenced by a specific subset of friends and close associates. His model is motivated by situations where word of mouth, or communication through a social network (as opposed to broadcasting or global visibility) is the dominant form of interaction. In such decentralized networks, spontaneous outbreaks of coordinated behavior can seem particularly mysterious. He poses the conundrum like this:

> Why do some books, movies and albums emerge out of obscurity, and with
> small marketing budgets, to become popular hits, when many a priori
> indistinguishable efforts fail to rise above the noise? Why does the stock market
> exhibit occasional large fluctuations that cannot be traced to the arrival of any
> correspondingly significant piece of information? How do large, grassroots
> social movements start in the absence of centralized control or public
> communication?

All these social phenomena involve herd behavior, where each person relies on the decisions of others to guide his or her own actions. More abstractly, imagine a network of any kind of nodes—companies, people, countries, or other decision makers—and suppose that each node is facing the same binary choice: adopt a new technology or not, riot or not, sign the Kyoto treaty or not. As in Granovetter's model, the decision to adopt, riot, or sign is determined by how many other nodes have already chosen to do so, except that now each node only pays attention to its specific set of "neighbors"—the nodes whose decisions influence it. (For example, a company's decision to buy a fax machine back in 1985, when they still seemed exotic, may have been strongly affected by whether its business partners had already done so, since fax machines became increasingly useful with the more contacts who had one.) Each node's threshold is defined as the fraction of neighbors who must take action before it will. To allow for diversity in the population, Duncan assumed that some nodes are more adventurous than others, and also that some are better connected. In mathematical terms, this means that both the thresholds and the numbers of neighbors are distributed across the population. Finally, given its allotted number of neighbors, each node forges those links to members of the population chosen at random (not realistic, but the analysis is hard enough even with this approximation).

The game starts when one node is randomly chosen as a seed, an innovator who decides to take the plunge. Visualize it as a domino falling over. Then, one by one, in random order, each node looks at its neighbors and checks what proportion of them have toppled. If its threshold has been transgressed, it tips. Otherwise it stays upright. After each node has taken its turn, the process of checking and toppling begins anew. Some dominoes may have tipped in the first round (namely those neighbors of the seed whose thresholds were low enough to be toppled by it). They in turn can initiate secondary waves of toppling. But if the seed is poorly connected, or if its neighbors are a conservative lot with high thresholds, the trend may fizzle at the outset.

In this idealized universe, Duncan was able to determine the exact conditions under which an enormous cascade will be triggered by a single domino. He also managed to work out the likelihood and size of such cascades, and the risk factors that predispose the network to be more or less vulnerable to them. The conclusions are necessarily statistical in character; nothing can be said in advance about any particular simulation on the computer. The fine details of the outcome are different from run to run. They depend on the location of the seed, on how the thresholds are distributed across the population, and on how the connectivity varies from node to node. Still, some striking trends emerge that would not have been easy to anticipate by common sense.

The main result is that the model displays two distinct phase transitions, popularly known as tipping points. If the network is too sparsely connected, it fragments into tiny islands and cascades can't spread beyond any of them. At a higher, critical level of connectivity—the first tipping point—the islands abruptly link together into a giant mesh and global cascades become possible. An initial seed can now trigger an epidemic of change that ultimately infects much of the population. With further increases in connectivity, the cascades at first become even larger and more likely, as one might expect, but then—paradoxically—they become larger yet *rarer,* suddenly vanishing when the network exceeds a critical density of connections. This second tipping point arises because of a dilution effect: When a node has too many neighbors, each of them has too little influence to trigger a toppling on its own. (Remember that each node compares its threshold to the *fraction* of its neighbors that have tipped, not the absolute number. The more neighbors there are, the less impact any one of them has, in a fractional sense.)

Just before this second tipping point, the outcome is extremely unpredictable in much the same way that real fads are. The network can be perturbed by thousands of hopeful seeds, each of which provokes at most a disappointing ripple that quickly peters out. By this measure, the network appears highly stable and resistant to outside disturbances. Then another seed comes along, seemingly indistinguishable from the others before it, yet this one triggers a massive cascade. In other words, near this second tipping point, fads are rare but gigantic when they do occur.

Here's what's going on, intuitively. Lurking within the network is a connected subset of nodes that Duncan calls the vulnerable cluster. The geometric structure of this cluster—the way it percolates through the rest of the network—is what matters. In marketing language the vulnerable cluster is composed of "early adopters": not innovators themselves but nodes that are poised and ready to tip, if just one of their neighbors has already toppled. Close to the second tipping point, the vulnerable cluster is spindly and almost invisible—it occupies a very small percentage of the whole network—so the odds of igniting it with a random seed are small. But once ignited, it spreads a slow-burning fire to its neighbors, enough of which pass it on to their neighbors, continuing inexorably until the entire giant component (the vast, interconnected meshwork of nodes that dominates the system) is engulfed in flame. What's amazing about this is that nearly all the nodes in the giant component are not early adopters; they are a more stubborn bunch with higher thresholds, known in the marketing literature as the "early and late majority." Yet because the network is so densely connected near the second tipping point, a spark that happens to ignite the vulnerable cluster is able to create enough momentum to detonate nearly everyone else.

Duncan's model is obviously a caricature—it leaves out the richness of real social structure, and assumes all friendships carry equal weight and that all seeds are equally infectious—but even so, it mimics the features of real fads that seem most puzzling: their unpredictability, scarcity, and arbitrariness. In particular, the creeping advance of an improbable cascade near the second tipping point is reminiscent of a low-budget hit that starts out slowly and builds by word of mouth.

The model also makes testable predictions, not about single fads (which are inherently unpredictable, according to the theory), but about the statistics of many of them, viewed in aggregate. These statistical conclusions offer guidance about what interventions are most likely to trigger cascades. For example, the analysis suggests that heterogeneity in the population has mixed effects. A broader range of thresholds destabilizes the system, making it more susceptible to fads (essentially because there are more early adopters to provide kindling), whereas a broader range in connectivity (greater variability in the number of neighbors per node) tends to stabilize it. Also, cascades tend to start in different places near the model's two tipping points. Near the first one, when the network is still sparse and barely connected, cascades are most easily initiated at the hubs, the nodes with the most connections. Near the second tipping point, the few cascades that do occur are typically seeded at average nodes, inconspicuous nobodies, simply because there are so many more of them.

In contrast to fads, there's at least one form of human group behavior that you can count on every day: the maddening crush of traffic at rush hour. According to most projections, it's only going to get worse. By 2020 the typical commute in Los Angeles is expected to take twice as long as it did in the 1990s, with traffic crawling along at an average of 24 miles an hour. Various proposals for unclogging highways are being considered, such as road-use fees, improvements in mass transit systems, and separate highways for cars and trucks. In the meantime, physicists and complexity theorists are taking a fresh look at the dynamics that cause congestion in the first place. Their new models suggest that traffic is more complex and unpredictable than traditionally imagined, largely because of nonlinear interactions between drivers.

Although we don't normally think of it in these terms, traffic is a social phenomenon in the sense that one driver's behavior affects that of others nearby. If someone swerves in front of you, you'll need to brake suddenly, and your reaction could trigger a wave of further braking behind you, in the worst case leading to a catastrophic pileup. Even in less dramatic situations, every driver has the power to impose his whims on others around him, by tailgating, or weaving aggressively, or honking for no reason. In that sense, congested traffic raises the conflict found in all social dilemmas: self-interest versus the common good. Everyone has an incentive to be selfish—altruistic drivers don't get home as fast. On the other hand, rampant selfishness makes driving unpleasant for all of us, as when some buffoon tries to inch across a busy intersection and gets trapped in the middle, blocking the crosswise traffic and causing gridlock.

So it came as a surprise recently when a model of traffic flow predicted that such widespread ruthlessness could, under the right circumstances, lead to a state of crystalline harmony that's ideal for all. This self-organized state was discovered in 1998 by Dirk Helbing, a leader in the emerging field of traffic physics, and Bernardo Huberman, a complexity theorist who normally spends his time thinking about the Internet. They were simulating the dynamics of a realistic mix of hundreds of virtual cars and trucks traveling along a two-lane highway. Each vehicle obeyed certain reasonable rules: accelerate to an optimal safe speed, slow down to avoid colliding with a vehicle too close in front, switch lanes and try to pass it (if there's enough room), and so on. The artificial drivers were even endowed with erratic, humanlike qualities, such as an occasional random tendency to dawdle after changing lanes.

Helbing and Huberman computed the long-term traffic patterns under a variety of different conditions. When there were only a few vehicles on the road, all the cars sailed past the slower-moving trucks without ever decelerating, while the trucks lumbered along at their maximum safe speed of 55 miles an hour. At higher but still moderate densities of traffic, some unlucky cars found themselves trapped behind trucks for a long time, with no room to pass or switch lanes.

At a critical density of traffic—about 35 vehicles in each lane per mile of road—all the cars and trucks spontaneously synchronized, traveling down the highway like a solid block. Remarkably, out of pure competition, with no coordinator or central authority, a large group of selfish individuals ended up in a cooperative state that was optimal for all of them. (Adam Smith would approve.) This state was optimal in the sense that the flux of traffic was as high as it could be: The number of cars and trucks passing through a given stretch of highway per hour was maximized. It was also the safest way for traffic to flow, because the drivers had no opportunities to change lanes or pass (the maneuvers associated with most accidents). Helbing and Huberman tested their model against data taken from a two-lane Dutch highway and found evidence of the predicted state. At the critical density, the car speeds were at their most stable, as measured by their velocity fluctuations, and lane changing and passing were minimized. Unfortunately—and as the model also predicted—the crystalline state proved to be delicate. At densities just above critical, it melted into a disorganized liquid state, which created opportunities for passing again, leading to unsteady, stop-and-go traffic.

Helbing and Huberman suggest the use of computer-controlled stoplights at on-ramps to help keep the solid block intact. The lights would respond to instantaneous data collected from electronic sensor wires that the cars pass over. If the sensors detect a gap following a block of traffic passing an on-ramp, the light would turn green to allow more cars to stream onto the highway, filling the gaps to keep traffic in sync; when the block threatens to dissolve into a stop-and-go pattern, the light turns red again. That strategy would differ from the one currently used on the Long Island Expressway, for example, where the on-ramp traffic lights are timed according to a preset schedule. The new approach still wouldn't cure the jams at the peak of rush hour, but at intermediate densities it might help traffic flow more safely and smoothly.

A different form of synchronized traffic was discovered a couple of years earlier by Boris Kerner and Hubert Rehborn, physicists at DaimlerChrysler in Stuttgart, when they were analyzing data collected from sensors built into German autobahns. For densities between free-flowing traffic and complete jams, they found a strange, highly congested state in which all the cars abruptly slowed down to the same speed and stayed in their lanes, creeping forward as a unified mass. But unlike the synchronized state found by Helbing and Huberman, this one was not coordinated by slow-moving trucks. It occurred all on its own, in a population of cars only. The spontaneous slowdown seemed to occur in the vicinity of on-ramps, when an unusually large horde of cars squeezed onto an already busy highway during the morning rush hour. The sudden influx somehow condensed the neighboring traffic, in the same way that a mote of dust can help water vapor condense into a droplet.

But what was really peculiar about this state is that it lingered for two hours, long after the inflow from the ramp returned to normal. In other words, the pattern takes on a life of its own. It is self-sustaining. It even sends waves of congestion backward down the highway. The later drivers encountering these stop-and-go waves find them mystifying. Delays occur periodically for no apparent reason.

Computer simulations later demonstrated that the pattern is not maintained by overloading per se. After the burst on the ramp subsides, the subsequent traffic could just as well have flowed freely, even with the same set of drivers, at the same density of traffic. That more pleasant alternative is just as stable and self-sustaining. But the drivers can't collectively achieve it. They are trapped in one stable mode, unable to reach a better one. In that respect, synchronized traffic is like the spiral and scroll waves in the BZ reaction, or the pernicious rotating waves responsible for cardiac arrhythmias. Once established, these waves are hard to kill. For immediate relief, the traffic needs to be defibrillated.

Unfortunately, no such technology exists. What actually happened on that particular German autobahn, on the day the data were collected, was that the pulsating congestion dragged on until 9:30 A.M. By that hour the ramp flow had thinned out so much that the pattern could no longer feed itself. The synchronized state spontaneously dissolved and traffic began to flow freely again.

Although traffic synchronization is unintentional, most forms of mass human synchrony are deliberate. It delights us to dance and sing together, stomp our feet, do "the wave" at a football game. When everybody is trying to cooperate, however, the group behavior that actually emerges can still hold some surprises. Consider, for example, an audience clapping in unison. That phenomenon seems self-explanatory, which is why we've invoked it repeatedly as a metaphor for other kinds of sync. But when scientists finally got around to measuring it, they were startled by what they found.

In 1999, a team of physicists, all of Eastern European descent, went to concert halls in Romania and Hungary and recorded several audiences clapping at the ends of opera and theater performances. The recordings showed that the audiences clapped tumultuously at first, then spontaneously switched to thunderous, rhythmic applause at a slower tempo, and then relapsed into cacophony, swinging back and forth six or seven times between chaos and sync. To explore the process in more detail, Zoltan Néda and his graduate student Erzsébet Ravasz asked individual high school students to stand alone in a room and clap in two different ways. First, each student was asked to clap as he or she would after an outstanding performance. This style of applause was found to be fast and irregular, averaging four claps per second but with wide variations, both within individuals and across the population. Then the experimenters asked the students to pretend they were clapping in sync with an imaginary audience. Now the clapping slowed down to a stately two beats per second— half as fast as before, as if people were skipping every other beat—while also becoming much more precise, as if there were a strong, shared understanding about what the right tempo should be.

The behavior of an entire audience can then be explained in these terms. Because of cultural expectations, the audience members all know that they want to clap in unison. But some have inherently faster or slower intrinsic clap rates. To get in sync, everyone slows down to half the rate of individualistic applause, and the dispersion of frequencies tightens up (as found in the experiments on high school students). Now, as in the coupled oscillator models of Winfree and Kuramoto, when the dispersion of frequencies is sufficiently reduced, the system abruptly crosses a phase transition and sync breaks out spontaneously. The twist in all this—the part that no theorist ever imagined— is that the synchrony comes with a psychological price. Although the collective clap is thunderous, it occurs only half as often as the faster, more raucous kind of applause, with the inevitable consequence that the total amount of noise summed over time is less than it would have been during disorganized clapping. Somehow, the audience feels that the cumulative level of noise does not adequately convey their excitement, so they make more noise the only way they can—by speeding up. But now their frequency distribution broadens as well (since faster clapping is inherently sloppier, as the measurements showed). So the phase transition is crossed in the opposite direction, and the group crumbles back into chaos. In a sense, the audience is frustrated by a trade-off between optimal synchronization and optimal noise intensity. They can't have both at the same time.

The authors wryly note that these swings between chaos and synchrony never occurred during the giant Communist rallies they had to endure in their youth. Audiences listening to the speeches of the "great leader" would dutifully applaud in listless synchrony, with no desire to speed up into disorder.

Even in a form of group behavior as automatic as hand clapping, human psychology enters in subtle ways. Yet for now at least, all the models neglect the vagaries of human volition. They deliberately pretend that people act like robots, to see how much can be explained on that basis alone. In Duncan Watts's model of fads, people flip once their threshold is exceeded. In traffic models, drivers speed up or slow down as the local conditions demand, as if enslaved to a human version of cruise control. In models of artificial societies, genocidal tribesmen don't act up when the United Nations peacekeepers are there, but go on a killing spree as soon as the troops are pulled out.

It's precisely because the models are so dumbed down that their fidelity can be so unnerving. In many forms of pack behavior, people don't rely on their higher cognitive abilities. "In individuals, insanity is rare," said Nietzsche, "but in groups, parties, nations, and epochs it is the rule." Maybe this is part of what we find so appalling about the spectacle of Nazis goose-stepping. In the hands of totalitarian regimes, synchrony becomes a symbol of all that is subhuman. "He who joyfully marches to music rank and file, has already earned my contempt," said Einstein. "He has been given a large brain by mistake, since for him the spinal cord would surely suffice."

The irony is that sync is just as much a part of the most beautiful forms of human expression, in ballet, in music, even in the love shared by people whose hearts are in sync. The difference is that these are more supple forms of sync, not mindless, not rigid, not brutally monotonous. They embody the qualities that we like to think of as uniquely human—intelligence, sensitivity, and the togetherness that comes only through the highest kind of sympathy.

Along with synchronization to each other, we sometimes feel like we're in sync with the world around us. The clearest example is our entrainment to the spin of the Earth, to the daily cycle of light and darkness. But aside from circadian rhythms, there aren't many well-documented cases of human sync to the environment.

For example, spooky effects have been ascribed to the phases of the moon. According to folklore, more crimes occur when the moon is full (also more suicides, psychiatric admissions, drug overdoses, and dog bites). There are even some scientific papers purporting to give statistical evidence for the "lunar effect." But when the statistics are redone properly, the correlation with lunar phase always evaporates. To give just one example of the shoddy studies in this area, some authors have claimed that more car accidents occur during a full moon, but forgot to control for weekly or seasonal variations in their incidence. Accidents are more frequent on Friday and Saturday nights, on New Year's Eve and other holidays, and during the summer (all for obvious reasons), so if any of those occur disproportionately during the time period studied, the statistics will be skewed accordingly. Statisticians who have adjusted for such calendar effects have found that the full moon makes no significant difference. Across the board, whether for fertility or homicide rates, assassinations or natural disasters, one careful study after another has demonstrated that the full moon has no measurable effect on human affairs. Yet many sensible people—including police officers and emergency room staff—continue to believe otherwise.

The lunar myth exemplifies the gullible side of our desire to find order in the universe, and particularly to connect the rhythms of our own lives to those of the cosmos. The same impulse drives quack notions about astrology and "biorhythms," a now-forgotten pseudoscience fashionable in the 1970s. (Back then you could buy a Casio watch equipped with a nifty biorhythm calculator, so you would know if you were about to have a bad day.) The theory claimed that our bodies are buffeted by predictable tides of physical ability, emotional condition, and intellectual performance, waxing and waning with periods of exactly 23, 28, and 33 days, supposedly the same for everyone regardless of age, sex, health, or genetic variability. In dozens of rigorous, independent studies by the military and airline industry in the 1970s, no evidence was found for any such biorhythms. Nor has any ever been found for Carl Jung's idea of "synchronicity," the claim that meaningful coincidences in our lives occur more often than one could explain by chance alone. Still, it's fun to believe in such things. In my own life, I've often wondered what made me wander into Heffer's bookstore that rainy day in England, and see a book with the peculiar title *The Geometry of Biological Time,* when just a year earlier I'd written a senior thesis whose subtitle was uncannily similar: "An Essay in Geometric Biology." Without that chance encounter with Art Winfree's book, and the fluky choice of the same words, I might never have met him, never gotten interested in sync, and never written the book you hold before you.

The problem with arguments like this is that all human beings—professional mathematicians included—are easily muddled when it comes to estimating the probabilities of rare events. Even figuring out the right question to ask can be confusing. In an article on coincidences, statisticians Persi Diaconis and Frederick Mosteller discuss the seemingly amazing case of a woman who won the New Jersey Lottery twice. A front-page story in the *New York Times* described that coincidence as a 1 in 17 trillion long shot, but that's calculating the right answer to the wrong question. This number assumes that the woman bought one ticket for exactly two lotteries, both of which were winners; in fact, she played the lottery frequently, and usually bought several tickets. The more relevant question is, What are the odds that with all the millions of people playing the lottery every day, year after year, that someone would hit it twice in a lifetime? When the question is framed that way, an event that once seemed astronomically unlikely is now exposed as a virtual certainty: The odds are better than 50-50 that over a period of just seven years, someone somewhere in the United States will win the lottery twice. To be fair, the New Jersey woman's luck was even better than that: She hit the jackpot twice in a four-month span. Even so, the odds of that happening to someone somewhere are better than 1 in 30: improbable, but not impossible.

The best case that can be made for human sync to the environment (outside of circadian entrainment) has to do with the possibility that electrical rhythms in our brains can be influenced by external signals. For instance, Norbert Wiener described an outrageous experiment conducted in Germany in the 1950s, in which the unnamed scientists attempted to synchronize a human subject's brain waves by beaming high-power electromagnetic radiation at him. As Wiener tells it, a sheet of tin was suspended from the ceiling and connected to one terminal of a 400-volt generator running at 10 cycles a second, the same frequency as the brain's alpha rhythm. He writes that this apparatus "can produce electrostatic induction in anything in the room" and that "it can actually drive the brain, causing a decidedly unpleasant sensation."

That sensation may have been something like what accidentally happened to hundreds of Japanese children watching an episode of *Pokémon* (pocket monsters) on the night of December 16, 1997. The hyperkinetic cartoon—the highest-rated television show in its 6:30 time slot—featured a scene in which a character destroyed a computer virus by detonating a "vaccine bomb." Viewers were subjected to a bright-white explosion followed by brilliant red, white, and blue lights that flashed like a strobe, 12 times a second, for five seconds. Kids around the country immediately began feeling sick. Some vomited. Others had seizures. A few stopped breathing momentarily. Their horrified parents lit up the phone banks of emergency services around the country, and more than 600 children were rushed by ambulance to emergency rooms. The attacks may have been exacerbated by the viewing conditions in Japanese homes, many of which are small and have large-screen televisions: Watching television in a Japanese apartment is like sitting in the front row of a movie theater. One fourteen-year-

old boy who was sitting less than three feet from his big-screen TV fell unconscious for more than a half hour. Even more people were stricken later that
night when Japanese news programs irresponsibly replayed excerpts from the
sickening scenes.

The intense optical stimulation caused by the pulsing, kaleidoscopic bursts
of light apparently triggered attacks of photosensitive epilepsy, a rare disorder
that has become much more common as television and video games have prolif-
erated. The precise cause of photosensitive epilepsy is unknown, but it's thought
to be a synchronization disorder in which brain waves are entrained by flickering
light, causing neurons in the brain to misfire in lockstep and produce a seizure.
That hypothesis is consistent with the clinical observation that the most dangerous frequencies are between 15 and 20 cycles a second, just a bit faster than the
brain's alpha rhythm. Here, then, is a case where a fast, periodic signal coming
from the external environment has a pronounced effect on human biology.

A more fleeting kind of sync appears to be implicated in one of the greatest
unsolved problems in human psychology: the mystery of how the brain gives
rise to the mind. Although scientists are still struggling to understand the neural basis of human thoughts and feelings, it has recently become possible to
eavesdrop on the mind as it recognizes a face, remembers a word, or snaps to
attention. Neurobiologists have discovered that such acts of cognition are
linked to a brief surge of neural synchrony, in which millions of far-flung brain
cells suddenly switch on and off in precise lockstep at about 40 times a second,
and then just as rapidly unravel to allow the next thought or perception to
occur. If this view is right, a flash of insight is literally a burst of electrical synchrony, an instant when separate parts of the brain begin to harmonize.

This line of research can be traced to the early 1980s, when Christoph von
der Malsburg of the University of Southern California proposed that neural
sync might provide a mechanism for solving the "binding problem," a long-
standing puzzle in brain science. To illustrate the problem, imagine that you are
sitting in a crowded, smoky café, sipping coffee and listening to rock music, as
people squeeze past your table and shout hellos to each other. With no effort at

all, you instantly perceive that you are holding a cup of coffee in your hand. But how, exactly, do you manage that? Simple as it seems, that perception is associated with a plethora of sensations. As you glance down at the coffee cup, light scatters off its surface and strikes your retinas, revealing its round shape, smooth texture, and white color. Each of those visual attributes is then sent to separate parts of your brain for further processing and interpretation. At the same time, vaporized coffee molecules bind to receptors in your nose, and trigger rhythmic bursts of neural activity in your olfactory centers (plus an additional burst of pleasure in your limbic system, associated with the sumptuous aroma of freshly ground beans). Meanwhile, other less desirable sensations— the smell of cigarette smoke, the jostling of people bumping your table as they slide past—are impinging on your senses as well, and exciting their own sets of neurons. The question is, How does your brain make sense of all this neural commotion? In particular, what physical process "binds" the right features together to form a unified perception of a cup, as distinct from the sound of the rock music, the shaking of the table, and all the other confusing sensations that are occurring simultaneously but are unrelated to it?

Von der Malsburg hypothesized that the separate banks of neurons processing the various features of the cup would all oscillate in sync for a fraction of a second. Their temporal coincidence would be the brain's way of binding them together, of signifying that they all refer to the same object. But he despaired of ever testing the idea. Even if the neural clusters did fire in concert, he supposed they would be drowned out by the incessant chatter of the brain's other electrical activity. "There would be no way to pick them out," he once said. "The mind would be invisible."

That pessimism turned out to be unwarranted. By 1989, glimpses of synchrony started to appear in experiments on animals. A team of neuroscientists led by Charles Gray and Wolf Singer showed an anesthetized cat an image of a moving bar, and found that the neurons responding to the bar began to fire rhythmic discharges at 30 to 60 cycles per second. The fusillade was short-lived, lasting about one-third of a second, but highly synchronized, with neurons hitting a series of corresponding electrical peaks and valleys along the way.

Perhaps most surprisingly, even cells that were separated by anatomically huge distances, halfway across the cat's visual cortex, managed to oscillate in nearly perfect unison. To test whether the coordinated firing meant the cat was perceiving the bar as a unified whole, Gray and Singer deleted the middle of the bar and moved both ends, giving it the appearance of two independent objects. The same brain cells continued to discharge but now fell out of step, just as Von der Malsburg would have predicted.

At the time, these findings provoked a storm of controversy. The air was filled with the usual arguments that always confront spectacular claims of synchrony. The most dubious skeptics denied the existence of the phenomenon, claiming the statistical analyses were erroneous, or that the transient correlation between distant neurons could have been produced by chance. Others fretted about the lack of any known mechanism that would allow neurons so far apart to synchronize as precisely as Gray and Singer were reporting. (It was hard to understand how cells could fire within a thousandth of a second of each other, despite being so widely separated that no neural impulse could travel between them in that time.) Over the next several years, however, these and other objections were cleared up, leaving only the concern that the sync might be real but meaningless, a useless by-product of the electrical activity in the cat's brain, no more revealing of its innermost workings than the 60-cycle electrical hum of a desktop computer.

Throughout the 1990s, the evidence linking synchrony to cognition became more persuasive. In experiments on animals ranging from locusts to monkeys, researchers found that synchronized neural activity is consistently associated with primitive forms of cognition, memory, and perception (for example, the ability to discriminate between two odors, or to detect a change in the orientation of a shape). But since it's impossible to know exactly what an animal is perceiving, the skeptics remained unconvinced. They wanted to see proof that synchrony was essential to cognition and not merely associated with it. In Gray and Singer's experiment, for example, there was no proof that the cats were perceiving a single bar in one case and two bars in the other, even though a person would see it that way. The only way to settle the issue was to perform experiments on human subjects.

One such study, reported in 2001 by Jürgen Fell and his colleagues at the University of Bonn in Germany, uncovered a tantalizing connection between neural synchrony and short-term memory. They asked volunteers to memorize lists of words, and after briefly distracting them with another task, tested their recall. Meanwhile, during the memorization phase of the experiment, the scientists measured the firing patterns of neurons in the subject's hippocampus and rhinal cortex, adjacent brain areas known to be involved in memory. (This experiment was remarkable in a technical sense, in that neural activity was measured directly, not inferred from brain waves. These subjects were epileptics who already had electrodes implanted in their brains in preparation for upcoming neurosurgical procedures, which afforded an unusual opportunity to record directly from human brain cells during the act of memorization.)

Naturally, each subject remembered some words and forgot others, but what was fascinating is that their neurons behaved differently in the two cases, at the moment the words were first viewed. A quarter of a second after viewing words that they'd later remember, their brains showed a rush of synchrony between the hippocampus and rhinal cortex, but there was no synchrony when they first viewed words that they'd later forget. To exaggerate a bit, this means that by watching the electrical pattern in someone's brain when he or she tries to memorize a word, you can predict whether he or she will succeed. You can see if the brain is dropping the ball.

It's unclear what to make of this surge of synchrony. It might be nothing more than the echo of a memory being formed by other, more important processes waiting to be discovered—just as thunder is an aftershock of lightning and not its cause. On the other hand, perhaps the synchrony is crucial to the memory process itself, as it would be if the chemical and electrical events associated with it somehow primed the hippocampus to store a new item or rendered that item more easily retrievable. That possibility is plausible on biological grounds; it is known that the connections between neurons are strengthened when they fire simultaneously, a principle often summarized as "neurons that fire together, wire together." By sparking tighter connections between neurons in critical brain areas, synchrony might pave the way for short-term memories to be laid down. Another possibility, and one that is always implicit when sync occurs, is that by firing in unison, the neurons stand out above the background chatter, just as people singing in unison would be audible above the din at a cocktail party. By coordinating their electrical activity, the synchronous neurons would amplify their message, making it more salient to the neurons downstream.

An even more intriguing experiment recently shed light on the puzzle of perception: how we pull the world together in our minds, and effortlessly integrate diverse sensations into coherent wholes. Some neurological patients are unable to do this, resulting in bizarre pathologies like the one famously described by Oliver Sacks in the title case of his book *The Man Who Mistook His Wife for a Hat*. The man was able to recognize her eyes, nose, mouth, and other parts of her face, but he couldn't put them together to see a whole face. For him, recognizing a face was an almost impossible task, and one which required the full force of conscious effort, whereas most of us do it instantly and unconsciously. The question is, What is going on in our brains when a face is recognized as a face, and not as a collection of unrelated parts?

In a 1999 study, a team of neuroscientists led by Francisco Varela asked volunteers to look at "Mooney faces," ambiguous black-and-white images that look like faces when they are viewed upright, but become meaningless blobs when viewed upside down.

The experimenters displayed one of these images on a computer screen and asked the subject to press one of two buttons as quickly as possible, depending on whether he or she perceived a face. Meanwhile the subject's brain waves were monitored through an array of 30 electrodes attached to his or her scalp.

About a quarter of a second after scrutinizing a picture, the subject's brain waves displayed a flurry of "gamma oscillations" caused by millions of neurons firing rhythmically at around 40 cycles a second in various regions of the cortex known to be associated with visual processing. These collective oscillations occurred in both cases, whether the image was a face or a blob. They apparently mark the moment of perception, the unconscious *Aha!* moment when the mind

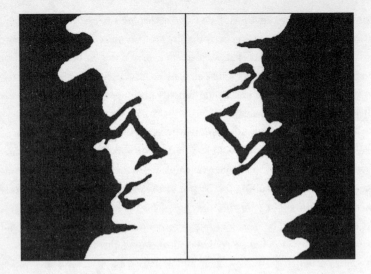

figures out what it is seeing. But although the rates of firing were similar in both conditions, the degree of synchrony was radically different. Only when a face was viewed did the electrical discharges align themselves in far-flung parts of the brain.

The distinction here is the same as that between a cacophony and a chorus. When perceiving a blob, the neurons in the various visual centers all sang in the same key of 40 cycles per second, but their timing was hopelessly off and the result was a meaningless racket, consistent with the brain's inability to make sense of the shape. On the other hand, when perceiving a face the neurons not only sang in the same key, but also in perfect time, suggesting that disparate features were being pulled together into an integrated perception of a face.

The subsequent events in the brain were equally fascinating. Even before the subject had a chance to react consciously and press a button, the surge of sync dissolved. The electrical coherence between neurons actively unraveled itself, like soldiers deliberately breaking step before marching across a bridge. Varela and his colleagues speculate that this active desynchronization may be the brain's way of wiping the slate clean, to allow another neuronal choir to form as the basis for the next thought or action. In this experiment, the subject's next action was to generate a motor response, to press a button on the computer. And, sure enough, about three-quarters of a second after an image was presented, corresponding closely to a typical subject's reaction time, the brain showed a second burst of sync, now between regions involved in the motor response. Not surprisingly, this second round of sync occurred whether a face was perceived or not, since a button was about to be pressed in either case.

These studies paint a disconcerting picture of human existence. As we go about our daily business, feeling in charge of our lives, we may be more robotic than we realize, clanking along from one neural state to another, feeling hungry, recognizing a friend's face, remembering to pick up milk on the way home, all depending on which banks of neurons happen to synchronize at any one moment. Some scientists have speculated that consciousness may be the subjective experience of these states of synchrony passing by in our brains. Others have gone even further, and suggested that sync may underlie consciousness itself. In a recent article titled "The Zombie Within," Caltech neuroscientist Christof Koch and his collaborator Francis Crick (the codiscoverer of the DNA double helix, and now a brain researcher at the Salk Institute) speculated that "consciousness involves synchronized firing of neurons at the millisecond level, whereas uncorrelated firing can influence behavior without generating that special buzz in the head."

At its strangest, that special buzz can trigger an indescribably odd sensation. If you have not felt it yourself, it will sound ridiculous, but if you have, you'll know exactly what I mean, and it's a chilling feeling. It happens to me maybe once or twice a year, and it comes without warning. I'll be standing in front of the mirror, brushing my teeth, and I look at myself and suddenly think: Who's in there? Or, Who's that?

I'm groping here for the words to express how odd it is to think about your own consciousness, your own self-awareness. In those weird moments in front of the mirror, I feel how strange it is to be conscious. Here is a pile of atoms, it looks like me, but I know it is a lot of water molecules, proteins, lipids, and all the rest, assembled in a particular way, and the damn thing is aware of itself and staring back at me. How does chemistry account for that, for me, for my feeling of identity? In other words, What is the physical foundation of consciousness?

No one knows yet, but it would be poetic justice if Koch and Crick turned out to be right. For if consciousness is the by-product of some sort of neural sync, then just thinking about sync—as you have been doing for a few hundred pages now—involves a stupendous act of sync itself.

EPILOGUE

I HOPE I'VE GIVEN YOU A SENSE of how thrilling it is to be a scientist right now. It feels like the dawn of a new era. After centuries of studying nature by teasing it into smaller and smaller pieces, we're starting to ask how to put the pieces back together again.

Old-timers will chuckle and say they've heard this line before. Every decade or so, a grandiose theory comes along, bearing similar aspirations and often brandishing an ominous-sounding C-name. In the 1960s it was cybernetics. In the '70s it was catastrophe theory. Then came chaos theory in the '80s and complexity theory in the '90s. In each case, the skeptics at the time grumbled that these theories were being oversold and that the results were either wrong or obvious. Then everyone had a good laugh and went back to the lab bench for more grinding, reductionistic science, walled off from their colleagues in adjoining disciplines, who were themselves grinding away on their own tiny corners of the universe.

What's different now is a feeling in the air. Even the most hard-boiled, mainstream scientists are beginning to acknowledge that reductionism may not be powerful enough to solve all the great mysteries we're facing: cancer, consciousness, the origin of life, the resilience of the ecosystem, AIDS, global warming, the functioning of a cell, the ebb and flow of the economy. It's a sign of the times, for example, that at every major research university, institutes are springing up with names such as functional genomics and integrative biology, where biologists are teaming up with computer scientists and mathematicians to try to make sense of the dance of life at the molecular level. Sequencing the human genome gave us an enormous list of parts: 30,000 individual genes and the proteins they encode. But we still have almost no clue how the interlocking activities of those genes and proteins are choreographed in the living cell.

What makes all these unsolved problems so vexing is their decentralized, dynamic character, in which enormous numbers of components keep changing their state from moment to moment, looping back on one another in ways that can't be studied by examining any one part in isolation. In such cases, the whole is surely not equal to the sum of the parts. These phenomena, like most others in the universe, are fundamentally *nonlinear*.

That's why nonlinear dynamics is central to the future of science. Chaos theory revealed that simple nonlinear systems could behave in extremely complicated ways, and showed us how to understand them with pictures instead of equations. Complexity theory taught us that many simple units interacting according to simple rules could generate unexpected order. But where complexity theory has largely failed is in explaining *where* the order comes from, in a deep mathematical sense, and in tying the theory to real phenomena in a convincing way. For these reasons, it has had little impact on the thinking of most mathematicians and scientists.

Here, it seems to me, is where sync has been uniquely successful. As one of the oldest and most elementary parts of nonlinear science (dealing, as it does, with purely rhythmic units), sync has offered penetrating insights into everything from cardiac arrhythmias to superconductivity, from sleep cycles to the stability of the power grid. It is grounded in rigorous mathematical ideas; it has passed the test of experiment; and it describes and unifies a remarkably wide range of cooperative behavior in living and nonliving matter, at every scale of length from the subatomic to the cosmic. Aside from its importance and intrinsic fascination, I believe that sync also provides a crucial first step for what's coming next in the study of complex nonlinear systems, where the oscillators are eventually going to be replaced by genes and cells, companies and people.

On the other hand, I don't want to leave you with a false impression. Sync is just a small part of a much larger body of thought. It is by no means the only approach to the study of complex systems. The chemist Ilya Prigogine and his colleagues feel that the key to unlocking the mysteries of self-organization lies in a deeper understanding of thermodynamics. They see the emergence of order as a victorious uphill battle against entropy, as a complex system feeds itself on energy flowing in from the environment. The community of physicists interested in pattern formation sees fluid mechanics as its paradigm, where the roiling of a turbulent fluid intermittently gives birth to coherent structures like helices and plumes, rather than degenerating into a bland, uniform smear. The physicist Hermann Haken and his colleagues view the world as a laser, with randomness and positive feedback conspiring to produce the organized forms that occur all around us. Researchers at the Santa Fe Institute are struck by the ubiquity of evolution through natural selection, not only in biological populations, but in immune systems, economies, and stock markets. Others conceive the universe to be a giant computer, running a cryptic program whose discovery would constitute the end of science.

But for now, these are mostly pipe dreams. We're still waiting for a major breakthrough in understanding, and it could be a long time in coming. I think we may be missing the conceptual equivalent of calculus, a way of seeing the consequences of the myriad interactions that define a complex system. It could even be that this ultracalculus, if it were handed to us, would be forever beyond human comprehension. We just don't know.

In the meantime, the science of synchrony is inching forward, one small step at a time. Charlie Peskin has started exploring the mechanics of flapping flight in insects. He's also refining his computer models of blood flow in the heart with his colleague David McQueen. Their simulations have already helped doctors design better artificial valves.

Yoshiki Kuramoto is close to retirement, but he is still blazing new trails.

He has been working hard on the mathematics of oscillators coupled in an intermediate way, not globally as in his classic model, but also not purely locally.

Chuck Czeisler is always in the news with important results about human sleep and circadian rhythms. He and his colleagues recently refuted an earlier and much ballyhooed study purporting to show that bright light applied to the back of the knee could reset the human circadian pacemaker. A year or two before that, NASA asked him to study John Glenn's circadian rhythms during his nostalgic flight on the space shuttle, to provide information about how aging affects the sleep-wake cycle.

Brian Josephson is still standing apart from the physics establishment, and busily updating his Web site with the latest news about homeopathy and paranormal phenomena. His former teacher Philip Anderson, now retired but active as ever, has spent more than a decade trying to crack the riddle of high-temperature superconductivity.

Kurt Wiesenfeld and his colleagues made a splash by redoing Huygens's pendulum clock experiment with modern equipment, and by using nonlinear dynamics to explain why the pendulums always end up swinging in perfect opposition.

Ed Lorenz was honored at a big international conference on complex systems in the spring of 2002, and, as usual, he said nothing in his lecture about his seminal work of 1963. "That little model" again took a backseat to what he is working on now, in his ninth decade of life.

Lou Pecora has been looking at synchronization in arrays of chaotic systems. He recently joined forces with one of my former students, Mauricio Barahona, to show that small-world networks are extraordinarily effective at synchronizing chaos, outperforming virtually all other kinds of architectures.

Duncan Watts is conducting an E-mail version of Milgram's small-world experiment, and László Barabási is pursuing the biological implications of scale-free networks.

Tragically, Art Winfree died on November 5, 2002, at age 60, seven months after being diagnosed with brain cancer. He helped me with this book at every stage, even when he was conscious only for a few hours a day. Though he did not live to see it published, he knew that it would be dedicated to him.

For many reasons, I'm not sure what I'm going to do next. There are so many problems to choose from. My students and I will certainly be studying some sort of group behavior in a complex nonlinear system, perhaps in connection with the gene networks that regulate the growth and division of cells, and which go haywire in cancer. The time seems ripe, given the explosive advances in our knowledge of biochemical networks, new technologies for tracking which genes are active at a given time, ever-increasing computer power, and the recent developments in network theory. It's too early to tell whether my favorite tools (idealized mathematical models and their qualitative analysis) will be too crude to shed any light on this agonizingly complicated and important set of questions. Experience has shown, however, that insisting on simplicity can help a lot, especially on problems where more realistic approaches can become tangled in a thicket of data. It's even possible that ideas from sync could be useful here, since cells act somewhat like oscillators, growing and dividing on a fairly regular cycle.

In any case, I'm sure that throughout my career, I'll keep returning to sync in one form or another. I find it beautiful and strange and profoundly moving, in a way that can only be described as religious. And I know I'm not alone in that reaction. When I read the old accounts written by sixteenth-century voyagers to Malaysia and Thailand, the first Westerners to witness the astonishing spectacle of fireflies flashing in unison for miles along the riverbanks, I hear in them that same sense of rapture. They all describe the displays with the same voice, stricken with such awe that later scientists found their reports easy to dismiss as unreliable and overly emotional.

For reasons I wish I understood, the spectacle of sync strikes a chord in us, somewhere deep in our souls. It's a wonderful and terrifying thing. Unlike many other phenomena, the witnessing of it touches people at a primal level. Maybe we instinctively realize that if we ever find the source of spontaneous order, we will have discovered the secret of the universe.

NOTES

CHAPTER 1 FIREFLIES AND THE INEVITABILITY OF SYNC

11   SO WROTE PHILIP LAURENT   Philip Laurent, "The supposed synchronal flashing of fireflies," *Science* 45 (1917), p. 44.

11   FOR 300 YEARS   An early account appears in the logs of Sir Francis Drake's 1577 expedition: "Our General . . . sayled to a certain little Island to the Southwards of Celebes . . . thoroughly growen with wood of a large and high growth. . . . Among these trees night by night, through the whole land, did shew themselves an infinite swarme of fiery wormes flying in the ayre, whose bodies beeing no bigger than our common English flies, make such a shew of light, as if every twigge or tree had been a burning candle" [R. Hakluyt, 1589. *A Selection of the Principal Voyages, Traffiques and Discoveries of the English Nation*. Edited by Laurence Irving (New York: Knopf, 1926), p. 151]. The synchronous aspect of the flashing was described much more explicitly in 1680 by the Dutch physician Engelbert Kaempfer, after a voyage down the Meinam River from Bangkok to the sea: "The Glowworms . . . represent another shew, which settle on some Trees, like a fiery cloud, with this surprising circumstance, that a whole swarm of these insects, having taken possession of one Tree, and spread themselves over its branches, sometimes hide their Light all at once, and a moment after make it appear again with the utmost regularity and exactness, as if they were in perpetual Systole and Diastole." [Engelbert Kaempfer, 1727. *The History of Japan (With a Description of the Kingdom of Siam)*. Translated by J. G.

Scheuchzer. London: Hans Sloane. Two volumes in one. See volume 1, p. 45, or pp. 78–79 of volume 1 of 1906 reprint by J. McLehose and Sons, Glasgow.]

11   20 OTHER ARTICLES   Many are cited in John B. Buck, "Synchronous rhythmic flashing of fireflies," *Quarterly Review of Biology* 13 (1938), pp. 301–314. This article is the best guide to the early literature on the subject.

11   THERE MUST BE A MAESTRO   George H. Hudson, "Concerted flashing of fireflies," *Science* 48 (1918), pp. 573–575.

12   "EXPLANATIONS ARE MORE REMARKABLE THAN THE PHENOMENON ITSELF"   Hugh M. Smith, "Synchronous flashing of fireflies," *Science* 82 (1935), pp. 151–152. In this brief but authoritative note, Smith also gave one of the most detailed descriptions of the phenomenon: "Imagine a tree thirty-five to forty feet high, thickly covered with small ovate leaves, apparently with a firefly on every leaf and all the leaves flashing in perfect unison at the rate of about three times in two seconds, the tree being in complete darkness between the flashes. . . . Imagine a tenth of a mile of river front with an unbroken line of Sonneratia [mangrove] trees with fireflies on every leaf flashing in synchronism, the insects on the trees at the end of the line acting in perfect unison with those between. Then, if one's imagination is sufficiently vivid, he may form some conception of this amazing spectacle."

12   AFRICAN VERSION   Joy Adamson, *Living Free* (London: Collins and Harvill, 1961). Quote from p. 29.

12   ELECTRICAL RHYTHM THAT TRAVELS DOWNSTREAM TO THE FIREFLY'S LANTERN   For more on the biochemistry underlying the flash rhythm, see Barry A. Trimmer et al., "Nitric oxide and the control of firefly flashing," *Science* 292 (2001), pp. 2486–2488.

13   DARKENED HOTEL ROOM   John Buck and Elisabeth Buck, "Mechanism of rhythmic synchronous flashing of fireflies," *Science* 159 (1968), pp. 1319–1327.

13   BUCK AND HIS COLLEAGUES   Frank E. Hanson, James F. Case, Elisabeth Buck, and John Buck, "Synchrony and flash entrainment in a New Guinea firefly," *Science* 174 (1971), pp. 161–164. A popular exposition of this and related work is given in John Buck and Elisabeth Buck, "Synchronous fireflies," *Scientific American* 234 (May 1976), pp. 74–85.

13   RESETTABLE OSCILLATOR   The resettable oscillator idea is discussed at length in John Buck, "Synchronous rhythmic flashing of fireflies. II," *Quarterly Review of Biology* 63 (1988), pp. 265–289, which appeared in the same journal, with the same title, exactly 50 years after his first review of the literature. This second review is still the definitive summary of what is known about firefly synchronization.

14   ONE OF THE MOST PERVASIVE DRIVES IN THE UNIVERSE   For an excellent, up-to-date review of the scientific and mathematical literature on synchronization, see Arkady Pikovsky, Michael Rosenblum, and Jurgen Kurths, *Synchronization: A Universal*

Concept in Nonlinear Science (Cambridge, England: Cambridge University Press, 2002).

14   SPERM SWIMMING   An early report of synchronized swimming of sperm appears in James Gray, *Ciliary Movement* (New York: Macmillan, 1928), especially Figure 78 on p. 119. See also G. I. Taylor, "Analysis of the swimming of microscopic organisms," *Proceedings of the Royal Society of London, Series A* 209 (1951), pp. 447–461. For the latest work explaining how the synchrony arises through mechanical forces conveyed by the fluid, see S. Gueron and K. Levit-Gurevich, "Computation of the internal forces in cilia: Application to ciliary motion, the effects of viscosity, and cilia interactions," *Biophysical Journal* 74 (1998), pp. 1658–1676.

15   PESKIN PROPOSED A SCHEMATIC MODEL   Charles S. Peskin, *Mathematical Aspects of Heart Physiology* (New York: Courant Institute of Mathematical Sciences Publication, 1975), pp. 268–278. Cardiologists now take a different view of how the pacemaker cells synchronize themselves. Peskin's model was predicated on the guess that synapses provide chemical coupling between pacemaker cells, whereas it is now thought that they are coupled electrically through gap junctions, which act like resistors. As such, the cells of the pacemaker are in constant electrical communication and interact throughout their cycle of activity, not only at the moment of firing, as Peskin assumed. For a more recent model, see D. C. Michaels, E. P. Matyas, and J. Jalife, "Mechanisms of sinoatrial pacemaker synchronization: A new hypothesis," *Circulation Research* 61 (1987), pp. 704–714.

19   FLIPPING THROUGH A BOOK   Arthur T. Winfree, *The Geometry of Biological Time* (New York: Springer-Verlag, 1980). The quote about Peskin's work is on p. 119. Winfree has recently updated his masterpiece (the second edition appeared in 2001), using a format that only he could think of, designed to highlight the twists and turns of scientific progress. Instead of taking advantage of 20 years of hindsight to repair the errors in the original, or to delete his own wrong guesses and predictions, he has chosen to leave the original intact, and to sequester new material in gray boxes around it, explicitly correcting or amplifying the old ideas (and in many cases, demonstrating how farsighted he actually was). Though disorienting at times, this unsanitized approach reveals science as the complicated, living, growing thing it truly is. (The effect is reminiscent of the marvelous "7 Up" series of documentaries by Michael Apted, in which people are interviewed every seven years throughout their lives, starting at age seven, and you get to see them at every stage as their lives unfold.)

23   WE WERE ABLE TO PROVE   Renato E. Mirollo and Steven H. Strogatz, "Synchronization of pulse-coupled biological oscillators," *SIAM (Society for Industrial and Applied Mathematics) Journal on Applied Mathematics* 50 (1990), pp. 1645–1662.

30   FIREFLIES THAT ARE MOST ADEPT AT SYNCHRONIZING For experimental documentation of the various resetting strategies used by fireflies, see Frank E. Hanson, "Comparative studies of firefly pacemakers," *Federation Proceedings* 37 (1978), 2158–2164. Our mathematical model was never intended to be realistic in this regard; we merely wanted to prove Peskin's conjecture, and cited fireflies as a vivid example of the abstract concept of pulse-coupled oscillators. A much more biologically faithful model of firefly synchrony is given in G. Bard Ermentrout, "An adaptive model for synchrony in the firefly *Pteroptyx malaccae*," *Journal of Mathematical Biology* 29 (1991), pp. 571–585.

30   IN NEUROBIOLOGY An early paper along these lines was L. F. Abbott and C. van Vreeswijk, "Asynchronous states in neural networks of pulse-coupled oscillators," *Physical Review E* 48 (1993), pp. 1483–1490.

30   HOPFIELD POINTED OUT A CONNECTION John J. Hopfield, "Neurons, dynamics, and computation," *Physics Today* 47 (1994), pp. 40–46; A.V.M. Herz and J. J. Hopfield, "Earthquake cycles and neural reverberations: Collective oscillations in systems with pulse-coupled threshold elements," *Physical Review Letters* 75 (1995), pp. 1222–1225.

31   SELF-ORGANIZED CRITICALITY For accessible accounts of self-organized criticality, see Per Bak, *How Nature Works: The Science of Self-Organized Criticality* (New York: Copernicus Books, 1999) and Mark Buchanan, *Ubiquity: The Science of History . . . or Why the World Is Simpler Than We Think* (New York: Crown, 2001).

31   DOZENS OF PAPERS For a review of the literature that links self-organized criticality to synchronization, see C. J. Pérez, A. Corral, A. Díaz-Guilera, K. Christensen, and A. Arenas, "On self-organized criticality and synchronization in lattice models of coupled dynamical systems," *International Journal of Modern Physics B* 10 (1996), pp. 1111–1151.

31   MEDIA ATTENTION See, for example: Ivars Peterson, "Step in time," *Science News* 140 (August 31, 1991), pp. 136–137; Ian Stewart, "All together now," *Nature* 350 (1991), p. 557; Walter Sullivan, "A mystery of nature: Mangroves full of fireflies blinking in unison," *New York Times* (August 13, 1991), p. C4.

33   WHAT WAS SO FAMILIAR TO THE FAUSTS The tale of Lynn Faust's discovery is told in Carl Zimmer, "Fireflies in lockstep," *Discover* 15 (June 1994), pp. 30–31, and in Susan Milius, "U.S. fireflies flashing in unison," *Science News* 155 (March 13, 1999), pp. 168–170. A charming firsthand account appeared in Lynn Faust, Andrew Moiseff, and Jonathan Copeland, "The night lights of Elkmont," *The Tennessee Conservationist* (May/June 1998), pp. 12–15. For the scientific documentation, see Andrew Moiseff and Jonathan Copeland, "Mechanisms of synchrony in the North American firefly *Photinus carolinus (Coleoptera: Lampyridae)*," *Journal of Insect Behavior* 8 (1995), p. 395.

33   NOT EVERYONE IS SO APPRECIATIVE   Dick Milne, "Govt. blows your tax $$ to study fireflies in Borneo: Not a bright idea!" *National Enquirer* (May 18, 1993), p. 23.

33   INTERNET ENGINEERS   Sally Floyd and Van Jacobson, "The synchronization of periodic routing messages," *IEEE-ACM Transactions on Networking* 2 (1994), pp. 122–136.

34   RESISTANT STRAINS OF TUBERCULOSIS   Anonymous, "Lighting the way: Tuberculosis sufferers are getting glowing help from the firefly," *Time* (May 17, 1993), p. 25. This article is based on the study by W. R. Jacobs et al., "Rapid assessment of drug susceptibilities of mycobacterium-tuberculosis by means of luciferase reporter phages," *Science* 260 (1993), pp. 819–822.

35   10 PLAUSIBLE EXPLANATIONS   The various hypotheses about the adaptive significance of firefly synchrony are summarized in John Buck, "Synchronous rhythmic flashing of fireflies. II," *Quarterly Review of Biology* 63 (1988), pp. 265–289.

35   LATEST THEORY   M. D. Greenfield and I. Roizen, "Katydid synchronous chorusing is an evolutionarily stable outcome of female choice," *Nature* 364 (1993), pp. 618–620. The idea that synchrony reflects competition was proposed here for katydids, but it might apply to fireflies, fiddler crabs, and other creatures as well.

35   PERIODICAL CICADAS   Susan Milius, "Cicada subtleties: What part of 10,000 cicadas screeching don't you understand?" *Science News* 157 (June 24, 2000), pp. 408–410. There has been a lot of intriguing speculation about why the reproductive cycles of cicadas are often 13 or 17 years, but never 12, 14, 15, 16, or 18 years. The explanation may have something to do with number theory. Both 13 and 17 are prime numbers (divisible only by themselves and 1), while the others are not. If potential predators have 2- to 5-year life cycles, as many of them probably do, this numerology helps the cicadas to avoid emerging in sync with the predators. See the chapter titled "Of bamboos, cicadas, and the economy of Adam Smith" in Stephen Jay Gould, *Ever Since Darwin: Reflections in Natural History* (Penguin Books, 1977). For an alternative theory, and a review of the recent literature on the cicada problem, see Eric Goles, Oliver Schulz, and Mario Markus, "Prime number selection of cycles in a predator-prey model," *Complexity* 6 (2001), pp. 33–38.

35   FIDDLER CRABS   P. R. Y. Backwell, M. D. Jennions, N. I. Passmore, and J. H. Christy, "Synchronous waving in a fiddler crab," *Nature* 391 (1998), pp. 31–32. A popular account appeared in Malcolm W. Browne, "Flirting male crabs found to wave claws in unison," *New York Times* (January 6, 1998), p. C4.

36   MENSTRUAL SYNCHRONY   The landmark paper is Martha K. McClintock, "Menstrual synchrony and suppression," *Nature* 229 (1971), pp. 244–245.

36   SOMETHING TO DO WITH PHEROMONES   Anonymous, "Olfactory synchrony of menstrual cycles," *Science News* 112 (July 2, 1977), p. 5. The original report was

published three years later in M. J. Russell, G. M. Switz, and K. Thompson, "Olfactory influences on the human menstrual cycle," *Pharmacology Biochemistry and Behavior* 13 (1980), pp. 737–738.

37   CHEMICAL COMMUNICATION BETWEEN WOMEN   Kathleen Stern and Martha K. McClintock, "Regulation of ovulation by human pheromones," *Nature* 392 (1998), pp. 177–179. McClintock's work on menstrual synchrony and human pheromones remains controversial. She gives a spirited defense of her work in Martha K. McClintock, "Whither menstrual synchrony?" *Annual Review of Sexual Research* 9 (1998), pp. 77–95. See also the entertaining and illuminating popular account given in Natalie Angier, *Woman: An Intimate Geography* (New York: Houghton Mifflin, 1999), pp. 170–175. She describes McClintock as "a woman of verve, rigor, and high, loopy enthusiasm who wears bright scarves over cashmere sweaters and unexpected accessories, like dove-gray socks patterned with black fishes."

CHAPTER 2   BRAIN WAVES AND THE CONDITIONS FOR SYNC

40   A UNIFIED FRAMEWORK   Norbert Wiener, *Cybernetics,* 2nd edition (Cambridge, Massachusetts: MIT Press, 1961).

40   WIENER WILL NEVER BE FORGOTTEN   For a survey of Wiener's work, and a small sample of the many hilarious anecdotes about him, see Pesi R. Masani, *Norbert Wiener 1894–1964* (Vita Mathematica, vol. 5), (New York: Springer-Verlag, 1990).

42   ALPHA RHYTHM   The final chapter of *Cybernetics* summarizes Wiener's ideas about the alpha rhythm of brain waves, and includes speculations about self-organization in other systems of coupled oscillators. (He thought it might have something to do with viruses, genes, and cancer.) For an earlier and more technical exposition, see Norbert Wiener, *Nonlinear Problems in Random Theory* (Cambridge, Massachusetts: MIT Press, 1958).

45   HE DREW A CARTOON VERSION   The double-dip spectrum is redrawn from a diagram on page 69 of Norbert Wiener, *Nonlinear Problems in Random Theory* (Cambridge, Massachusetts: MIT Press, 1958).

45   "WITHOUT DARING TO PRONOUNCE"   *Cybernetics*, page 201.

46   WHEN WINFREE THOUGHT ABOUT THE PROBLEM   His earliest work on group sync, in 1965, was based on an experiment involving an array of 71 flickering neon lamps coupled electrically to one another. Winfree called this gadget "the firefly machine," and wrote that his aim was "just to look and see what would happen"—see Chapter 11, *The Geometry of Biological Time*. He soon realized that computer simulation would provide much greater flexibility, control, and ease of interpretation. The results of those investigations are described in Arthur T. Winfree, "Biological

rhythms and the behavior of populations of coupled oscillators," *Journal of Theoretical Biology* 16 (1967), pp. 15–42, on which the rest of this section is based.

46   COMPLICATIONS THAT WOULD HAVE REPULSED NEARLY ANYONE ELSE   For readers with training in mathematics or physics: You may be wondering what was so novel about the problem that Winfree set for himself, and in particular, how it differed from what we are all taught about coupled oscillators. What you need to remember is that the textbook problems always assume that the oscillators are *linear* (that is, they are simple harmonic oscillators) coupled by *linear* interactions (e.g., by using springs that obey Hooke's law). In this simple case, the dynamics are explicitly solvable by the technique of normal modes. But Winfree realized that this would be irrelevant to the biological problem, because biological oscillators are not linear. Unlike their linear cousins, which can cycle at *any* amplitude, most biological oscillators stubbornly regulate their amplitude; hence, they are best modeled as nonlinear, self-sustained oscillators with stable limit cycles. In the mid-1960s, the available mathematical theory of such beasts ended at systems of two or three coupled limit-cycle oscillators. No one knew anything about *populations* of them, especially if their frequencies were randomly distributed across the population. Also, please realize that such oscillators should not be confused with conservative nonlinear oscillators (like the anharmonic oscillators used in molecular dynamics). These conserve energy and can have any amplitude—again, an inappropriate assumption for modeling biological, self-sustained oscillators.

49   CUT TO THE SIMPLEST PROBLEM   In the language of statistical physics, Winfree was making a "mean-field" approximation.

51   *NONLINEAR*   For an introduction to nonlinear differential equations, see Steven H. Strogatz, *Nonlinear Dynamics and Chaos: With Applications to Physics, Biology, Chemistry, and Engineering* (Cambridge, Massachusetts: Perseus Books, 1994).

55   KURAMOTO'S MODEL   The original paper—an almost impenetrably brief note—is Y. Kuramoto, "Self-entrainment of a population of coupled nonlinear oscillators," in *International Symposium on Mathematical Problems in Theoretical Physics*, edited by H. Araki (Springer-Verlag: Lecture Notes in Physics, vol. 39, 1975), pp. 420–422. A much clearer treatment is given in Y. Kuramoto, *Chemical Oscillations, Waves, and Turbulence* (Berlin: Springer-Verlag, 1984). For a pedagogical review of the model and its mathematical analysis, see Steven H. Strogatz, "From Kuramoto to Crawford: Exploring the onset of synchronization in populations of coupled oscillators," *Physica D* 143 (2000), pp. 1–20.

59   NANCY KOPELL   For an introduction to her work on coupled oscillators applied to neurobiology, see Nancy Kopell, "Toward a theory of modelling central pattern generators," in *Neural Control of Rhythmic Movement in Vertebrates*, edited by A. H. Cohen, S. Rossignol, and S. Grillner (New York: John Wiley, 1988), pp. 369–413.

298 · NOTES

61 "OSCILLATOR FLUID" Steven H. Strogatz and Renato E. Mirollo, "Stability of incoherence in a population of coupled oscillators," *Journal of Statistical Physics* 63 (1991), pp. 613–635.

64 STRANGE RESULTS Steven H. Strogatz, Renato E. Mirollo, and Paul C. Matthews, "Coupled nonlinear oscillators below the synchronization threshold: Relaxation by generalized Landau damping," *Physical Review Letters* 68 (1992), pp. 2730–2733.

64 LANDAU DAMPING Lev Landau, "On the vibrations of the electronic plasma," *Journal of Physics USSR* 10 (1946), pp. 25–34. For an elementary introduction, see David Sagan, "On the physics of Landau damping," *American Journal of Physics* 62 (1994), pp. 450–462.

64 CAR ACCIDENT Isaac Asimov, *Asimov's Biographical Encyclopedia of Science and Technology* (Garden City, New York: Doubleday, 1972), p. 723.

65 SOLVE A LONG-STANDING PROBLEM John David Crawford was a gracious and brilliant applied mathematician who died at a tragically young age after a battle with cancer. For a glimpse of his formidable work on coupled oscillators and plasmas, see John David Crawford, "Amplitude expansions for instabilities in populations of globally-coupled oscillators," *Journal of Statistical Physics* 74 (1994), pp. 1047–1084, and "Amplitude equations for electrostatic waves: Universal singular behavior in the limit of weak instability," *Physics of Plasmas* 2 (1995), pp. 97–128.

65 DO THEY PREDICT The first experimental test of the Kuramoto model was reported recently in a system of coupled chemical oscillators; see Istvan Z. Kiss, Yumei Zhai, and John L. Hudson, "Emerging coherence in a population of chemical oscillators," *Science* 296 (2002), pp. 1676–1678. Hudson and his colleagues verified the phase transition that Winfree and Kuramoto had predicted: Synchronization broke out abruptly once the coupling between the oscillators exceeded a certain threshold. They also found that the order parameter (the measure of how synchronized the oscillators are) grew as the coupling strength was increased, with the precise mathematical dependence between order and coupling that Kuramoto anticipated. But no comparably precise test has yet been reported for *biological* oscillators.

68 HE DESCRIBES THE SPECTRUM *Cybernetics*, pp. 190–191.

69 THE BRAIN DOES CONTAIN A POPULATION OF OSCILLATORS In all mammals, the master circadian clock is localized in a tiny pair of neural clusters situated just above the optic chiasm, the site where the optic nerves crisscross en route to the brain. The twin clusters, known as the suprachiasmatic nuclei, together contain thousands of specialized neurons that collectively generate an electrical signal which waxes and wanes on a 24-hour cycle, orchestrating the tissues and organs in the animal's body and coordinating their daily functions. Welsh and Reppert's new finding was that the individual cells are capable of spontaneous oscillation; even when they were removed from a rat's

brain and isolated from one another, they continued to fire electrical discharges for weeks. At some times of day they were silent; at other times they buzzed furiously. The disembodied cells continued to behave like responsible little alarm clocks, steadfastly ringing the wake-up call for an animal that no longer needed it. Furthermore, different cells had different natural periods, ranging from 20 to 25 hours. The distribution of periods was roughly bell-shaped, though its precise contour is not known yet. See D. K. Welsh, D. E. Logothetis, M. Mesiter, and S. M. Reppert, "Individual neurons dissociated from rat suprachiasmatic nucleus express independently phased circadian firing rhythms," *Neuron* 14 (1995), pp. 697–706.

Furthermore, Reppert and his colleagues showed in 1997 that mutant hamsters with fast clock cells, say with an average period of 20 hours, had correspondingly fast activity rhythms—they would jump onto the running wheels in their cages every 20 hours instead of every 24. To put it plainly, if your clock cells run fast, you'll run fast. Similar experiments on mice showed that the periods of an animal's clock cells are more broadly distributed than those of its behavioral rhythms. In other words, sloppy clocks conspire to make a more precise organism. That observation is consistent with Wiener's idea that the ensemble takes an average over the widely dispersed periods of its constituents, and therefore will be a more accurate clock than any one of them; see Chen Liu, David R. Weaver, Steven H. Strogatz, and Steven M. Reppert, "Cellular construction of a circadian clock: Period determination in the suprachiasmatic nuclei," *Cell* 91 (1997), pp. 855–860, and the related report by Erik D. Herzog, Joseph S. Takahashi, and Gene D. Block, "Clock controls circadian period in isolated suprachiasmatic nucleus neurons," *Nature Neuroscience* 1 (1998), pp. 708–713.

CHAPTER 3 SLEEP AND THE DAILY STRUGGLE FOR SYNC

71   INTERNAL BODY CLOCKS   Good general references about human sleep and circadian rhythms include Martin C. Moore-Ede, Frank M. Sulzman, and Charles A. Fuller, *The Clocks That Time Us: Physiology of the Human Circadian Timing System* (Cambridge, Massachusetts: Harvard University Press, 1982); Richard M. Coleman, *Wide Awake at 3:00 A.M.: By Choice or By Chance?* (New York: W. H. Freeman, 1986); Arthur T. Winfree, *The Timing of Biological Clocks* (New York: Scientific American Press, 1987).

71   "BEING BLIND IS OKAY"   Quoted in Lynne Lamberg, "Blind people often sleep poorly: Research shines light on therapy," *Journal of the American Medical Association* 280 (October 7, 1998), p. 1123.

72   ONE OF THE HOTTEST FIELDS   After 40 years of frustration, circadian biologists are finally beginning to figure out how circadian rhythms are generated at the molecular

level. For an unabashedly joyous review of these breakthroughs, see Steven M. Reppert, "A clockwork explosion!" *Neuron* 21 (1998), pp. 1–4. A more recent summary is Steven M. Reppert and David R. Weaver, "Molecular analysis of mammalian circadian rhythms," *Annual Review of Physiology* 63 (2001), pp. 647–676.

72 SUITES OF GENES Kai-Florian Storch et al., "Extensive and divergent circadian gene expression in liver and heart," *Nature* 417 (2002), pp. 78–83.

72 SYNCHRONY OCCURS BETWEEN THE VARIOUS ORGANS Shin Yamazaki et al., "Resetting central and peripheral circadian oscillators in transgenic rats," *Science* 288 (2000), pp. 682–685.

73 CRYPTIC REGULARITIES Steven H. Strogatz, *The Mathematical Structure of the Human Sleep-Wake Cycle* (Lecture Notes in Biomathematics, vol. 69) (New York: Springer-Verlag, 1986).

73 "A ROSETTA STONE" Arthur T. Winfree, "The tides of human consciousness: Descriptions and questions," *American Journal of Physiology* 245 (1982), pp. R163–R166.

74 TIME-ISOLATION EXPERIMENT Michel Siffre, "Six months alone in a cave," *National Geographic* 147 (March 1975), pp. 426–435.

75 OGLE FIRST REPORTED J. W. Ogle, "On the diurnal variations in the temperature of the human body in health," *St. George's Hospital Reports* 1 (1866), pp. 220–245. Quoted in Moore-Ede et al. (1982), p. 14.

76 INTERNAL DESYNCHRONIZATION J. Aschoff, "Circadian rhythms in man," *Science* 148 (1965), pp. 1427–1432. For a summary of the pioneering work of Aschoff and his collaborator Rutger Wever, see Wever's monograph *The Circadian System of Man* (Berlin: Springer-Verlag, 1979).

76 "JAGGED, SEEMINGLY RANDOM" Siffre (1975), p. 435.

78 ONE OF THE FORMER SUBJECTS RECALLED The quote is from Coleman (1986), p. 10. Coleman also provides other interesting details on what the experience was like in the Montefiore time-isolation facility.

79 OF THE FIRST 12 SUBJECTS C. A. Czeisler, E. D. Weitzman, M. C. Moore-Ede, J. C. Zimmerman, and R. S. Knauer, "Human sleep: Its duration and organization depend on its circadian phase," *Science* 210 (1980), pp. 1264–1267.

83 THE CLOUD WAS STRIKINGLY ASYMMETRICAL Its asymmetry was obscured in the original publication of Czeisler et al. (1980). The authors averaged the data at each phase before plotting it, giving it the misleading appearance of a sine wave. The raw data shown here were collected from a larger sample of subjects; see Steven H. Strogatz, Richard E. Kronauer, and Charles A. Czeisler, "Circadian regulation dominates homeostatic control of sleep length and prior wake length in humans," *Sleep* 9 (1986), pp. 353–364.

83   TRAIN DRIVERS   J. Foret and G. Lantin, "The sleep of train drivers: An example of
the effects of irregular work schedules on sleep," in *Aspects of Human Efficiency*,
edited by W. P. Colquhoun (London: English University Press, 1972), pp. 273–282.
The same paradoxical effect (after going to sleep later, you sleep less) was also documented in subjects who were living on a normal schedule, entrained in the usual way
to the 24-hour clock: T. Akerstedt and M. Gillberg, "The circadian variation of
experimentally displaced sleep," *Sleep* 4 (1981), pp. 159–169.

85   FIELD STUDIES SHOW   Moore-Ede et al. (1982), pp. 332–334.

85   RAPID-EYE MOVEMENT (REM) SLEEP   C. A. Czeisler, J. C. Zimmerman, J. Ronda,
M. C. Moore-Ede, and E. D. Weitzman, "Timing of REM sleep is coupled to the
circadian rhythm of body temperature in man," *Sleep* 2 (1980), pp. 329–346. See
also Czeisler et al. (1980); Moore-Ede et al. (1982), pp. 205–215; Coleman (1986),
pp. 104–130.

86   PUTTING THE CAR IN THE GARAGE   I'm not sure who first came up with this analogy. It's now part of the culture among sleep researchers. Philippa Gander used it in
her Cawthron Memorial Lecture, October 1997, "Sleep, Health, and Safety: Challenges in a 24-hour Society," available on-line at www.cawthron.org.nz/Assets/
Cawlec97.pdf.

88   PERIPHERAL CLOCKS   Early work in the field is reviewed in Moore-Ede et al.
(1982), pp. 134–139. For more recent developments in this rapidly moving branch
of circadian biology, see Yamazaki et al. (2000); Storch et al. (2002); P. McNamara
et al., "Regulation of CLOCK and MOP4 by nuclear hormone receptors in the vas-
culature: A humoral mechanism to reset a peripheral clock," *Cell* 105 (2001),
pp. 877–889; C. Schubert, "Vitamin A calibrates a heart clock, 24–7," *Science News*
160 (July 14, 2001), p. 22; and Michael H. Hastings, "A gut feeling for time,"
Nature 417 (2002), pp. 391–392.

88   GRUESOME SERIES OF EXPERIMENTS   For a review of Richter's work, and the later
work that localized the master clock in the suprachiasmatic nuclei, see Moore-Ede et
al. (1982), pp. 152–157.

89   THE DETAILS OF HOW THE PACEMAKER WORKS   Steven M. Reppert and David R.
Weaver, "Molecular analysis of mammalian circadian rhythms," *Annual Review of
Physiology* 63 (2001), pp. 647–676.

89   COUPLED PERHAPS BY CHEMICAL DIFFUSION   Chen Liu and Steven M. Reppert,
"GABA synchronizes clock cells within the suprachiasmatic circadian clock," *Neuron*
25 (2000), pp. 123–128.

90   CANCER CHEMOTHERAPY   F. Levi, "From circadian rhythms to cancer chronothera-
peutics," *Chronobiology International* 19 (2002), pp. 1–19; W. J. M. Hrushesky,
"Circadian timing of cancer chemotherapy," *Science* 228 (1985), pp. 73–75; W. J. M.

Hrushesky, "Tumor chronobiology," *Journal of Controlled Release* 74 (2001), pp. 27–30.

90   HEART ATTACKS   J. A. Panza, S. E. Epstein, and A. A. Quyyumi, "Circadian variation in vascular tone and its relation to alpha-sympathetic vasoconstrictor activity," *New England Journal of Medicine* 325 (1991), pp. 986–990; P. M. Ridker et al., "Circadian variation of acute myocardial-infarction and the effect of low-dose aspirin in a randomized trial of physicians," *Circulation* 82 (1990), pp. 897–902.

90   LIVE AN EXACT, WHOLE NUMBER OF DAYS   Moore-Ede et al. (1982), p. 348.

90   WINFREE WAS ESPECIALLY IMPRESSED   A. T. Winfree, "Human body clocks and the timing of sleep," *Nature* 297 (1982), pp. 23–27.

91   WINFREE KEPT HARPING   Arthur T. Winfree, "Circadian timing of sleepiness in man and woman," *American Journal of Physiology* 243 (1982), pp. R193–R204.

91   A HUGE DATABASE   Strogatz (1986), Chapter 3.

92   FORBIDDEN ZONES   Steven H. Strogatz, Richard E. Kronauer, and Charles A. Czeisler, "Circadian pacemaker interferes with sleep onset at specific times each day: Role in insomnia," *American Journal of Physiology* 253 (1987), pp. R172–R178. The Israeli sleep researcher Peretz Lavie independently discovered the forbidden zones at around the same time; see Peretz Lavie, "Ultrashort sleep-waking schedule. 3. Gates and forbidden zones for sleep," *Electroencephalography and Clinical Neurophysiology* 63 (1986), pp. 414–425. They have been further explored in, e.g., L. C. Lack and K. Lushington, "The rhythms of human sleep propensity and core body temperature," *Journal of Sleep Research* 5 (1996), pp. 1–11.

93   SIESTA TIME   For an early suggestion that afternoon napping might be built into our biology, see Roger Broughton, "Biorhythmic variations in consciousness and psychological functions," *Canadian Psychological Review* 16 (1975), pp. 217–239.

94   SINGLE-VEHICLE TRUCK ACCIDENTS   P. M. Lavie, M. Wollman, and I. Pollack, "Frequency of sleep-related traffic accidents and hour of the day," *Sleep Research* 15 (1986), p. 275. For a broader perspective, see M. M. Mitler et al., "Catastrophes, sleep and public policy: Consensus report," *Sleep* 11 (1988), pp. 100–109.

95   HOURLY DISTRIBUTION OF MICROSLEEPS   For a summary of Carskadon's data on unintended microsleeps during a constant routine, see Strogatz (1986), pp. 97–98.

95   "90-MINUTE DAY"   M. A. Carskadon and W. C. Dement, "Sleep studies on a 90-minute day," *Electroencephalography and Clinical Neurophysiology* 39 (1975), pp. 145–155; M. A. Carskadon and W. C. Dement, "Distribution of REM sleep on a 90-minute sleep-wake schedule," *Sleep* 2 (1980), pp. 309–317.

96   KRONAUER FOUND FURTHER EVIDENCE   J. E. Fookson et al., "Induction of insomnia on non-24 hour sleep-wake schedules," *Sleep Research* 13 (1984), p. 220. See Strogatz (1986), pp. 100–101, for the actual data.

97   "DELAYED SLEEP PHASE SYNDROME"   C. A. Czeisler et al., "Chronotherapy: Reset-
      ting the circadian clocks of patients with delayed sleep phase insomnia," *Sleep* 4
      (1981), pp. 1–21.

98   24-HOUR SOCIETY   Martin Moore-Ede, *The Twenty-Four-Hour Society: Under-
      standing Human Limits in a World That Never Stops* (Reading, Massachusetts:
      Addison-Wesley, 1993).

98   NUCLEAR SUBMARINES   T. L. Kelly et al., "Nonentrained circadian rhythms of
      melatonin in submariners scheduled to an 18-hour day," *Journal of Biological
      Rhythms* 14 (1999), pp. 190–196. The turnover data from submarine crews of the
      1970s are reviewed in Moore-Ede et al. (1982), pp. 336–337.

98   SUNLIGHT   For the first quantification of the effects of light on the human circa-
      dian pacemaker, see C. A. Czeisler et al., "Bright light induction of strong (type 0)
      resetting of the human circadian pacemaker," *Science* 244 (1989), pp. 1328–1333.
      These and later results are reviewed in C. A. Czeisler, "The effect of light on the
      human circadian pacemaker," *CIBA Foundation Symposia* 183 (1995),
      pp. 254–290.

99   NOT THE RODS AND CONES   M. Freedman et al., "Non-rod, non-cone photorecep-
      tors regulate the photoentrainment of locomotor behavior," *Science* 284 (1999),
      pp. 502–504; R. J. Lucas et al., "Non-rod, non-cone photoreceptors regulate the
      acute inhibition of pineal melatonin," *Science* 284 (1999), pp. 505–507.

99   BLIND PEOPLE   C. A. Czeisler et al., "Suppression of melatonin secretion in some
      blind patients by exposure to bright light," *New England Journal of Medicine* 332
      (1995), pp. 6–11; E. B. Klerman et al., "Nonphotic entrainment of the human cir-
      cadian pacemaker," *American Journal of Physiology* 43 (1998), pp. R991–R996.

100   "FAMILIAL ADVANCED SLEEP PHASE SYNDROME"   K. L. Toh et al., "An h*Per2* phos-
      phorylation site mutation in familial advanced sleep phase syndrome," *Science* 291
      (2001), pp. 1040–1043.

CHAPTER 4 THE SYMPATHETIC UNIVERSE

103   ANDROSTHENES   H. Bretzl, *Botanische Forschungen des Alexanderzuges* (Leipzig:
      B. G. Teubner, 1903), as cited in Martin C. Moore-Ede, Frank M. Sulzman, and
      Charles A. Fuller, *The Clocks That Time Us: Physiology of the Human Circadian Tim-
      ing System* (Cambridge, Massachusetts: Harvard University Press, 1982), p. 5.

103   HOW DIFFERENT SERENDIPITY IS FROM LUCK   My eyes were opened to this by R. S.
      Root-Bernstein's fascinating essay, "Setting the stage for discovery: Breakthroughs
      depend on more than luck," *The Sciences* 28 (1988), pp. 26–34. For further insights
      into the creative process, see Robert Root-Bernstein and Michele Root-Bernstein,

Sparks of Genius: The Thirteen Thinking Tools of the World's Most Creative People (Boston: Mariner Books/Houghton Mifflin, 1999).

104 "SLIGHT INDISPOSITION" C. Huygens, letter to R. Moray, dated February 27, 1665, in *Oeuvres Completes des Christian Huygens,* edited by M. Nijhoff (The Hague: Societé Hollandaise des Sciences, 1893), vol. 5, pp. 246–249. As he describes the sympathy of clocks to Moray, Huygens can barely contain himself: "This discovery thrilled me not a little . . ."

104 HUYGENS HAD INVENTED THE PENDULUM CLOCK C. Huygens, *The Pendulum Clock: Geometrical Demonstrations Concerning the Motion of Pendula as Applied to Clocks,* translated by R. J. Blackwell (Ames: Iowa State University Press, 1986). An on-line biography of Huygens can be found at http://www-history.mcs.st-and.ac.uk/history/Mathematicians/Huygens.html.

104 LONGITUDE For a captivating account of all aspects of the longitude problem, from science to political history to biography, see Dava Sobel, *Longitude: The True Story of a Lone Genius Who Solved the Greatest Scientific Problem of His Time* (New York: Walker Publishing Company, 1995).

106 IN A LETTER TO HIS FATHER C. Huygens, Letter to his father, dated February 26, 1665, in *Oeuvres Completes des Christian Huygens,* edited by M. Nijhoff (The Hague: Societé Hollandaise des Sciences, 1893), vol. 5, p. 243.

106 R. F. DE SLUSE C. Huygens, *Oeuvres Completes,* vol. 5, p. 241.

107 EACH CLOCK WAS HOUSED IN A HEAVY BOX A recent replication of Huygens's experiments, and the first explanation of the spontaneous synchrony in terms of nonlinear dynamics, is given in M. Bennett, M. F. Schatz, H. Rockwood, and K. Wiesenfeld, "Huygens's clocks," *Proceedings of the Royal Society of London, Series A: Mathematical, Physical, and Engineering Sciences* 458 (2002), pp. 563–579. For a popular account of this work, see Erica Klarreich, "Huygens's clocks revisited," *American Scientist* 90 (July/August 2002), pp. 322–323. This study reveals yet another layer of serendipity in Huygens's work. His design called for the clocks to be weighted with 80 or 90 pounds of lead to help them stay upright on the deck of a rolling ship, even when buffeted by stormy seas. The new analysis shows that if his clocks had been weighted with just a little more lead, they would have been too weakly coupled; the wooden support between them wouldn't have shaken enough for them to feel each other, and they wouldn't have synchronized. A little less lead, on the other hand, and they would have shaken each other so vigorously that one of them would have stopped swinging altogether (because at some point in its erratic motion, this pendulum would fall to such a low-amplitude swing that the clock's escapement mechanism would fail to engage, cutting off the energy supply needed

to keep the clock running). In other words, Huygens just happened to build his clocks in the narrow sliver of possible designs for which sync could occur.

109 LASERS Peter W. Milonni and Joseph H. Eberly, *Lasers* (New York: Wiley-Interscience, 1988). A good introduction to laser surgery is M. W. Berns, "Laser surgery," *Scientific American* 264 (June 1991), pp. 84–90.

109 AT A PARTY Schawlow is quoted at http://www.bell-labs.com/history/laser/today/sockhop7.html.

112 IT TOOK ANOTHER 43 YEARS The history surrounding the invention of the laser is tangled and controversial, involving a nasty battle between a Nobel laureate (Charles Townes) and a former graduate student named Gordon Gould, who recently won a court case giving him the patent rights. Townes is revered among scientists; he was the first to see how to apply Einstein's ideas about stimulated emission, leading to his 1954 creation of a device called a maser (the forerunner of the laser, which used microwaves instead of visible light). Townes gives his version of events in *How the Laser Happened: Adventures of a Scientist* (Oxford, England: Oxford University Press, 1999). For a dramatic telling of the story from Gould's point of view, see Nick Taylor, *Laser: The Inventor, the Nobel Laureate, and the Thirty-Year Patent War* (New York: Simon & Schuster, 2000). In any case, neither of them actually built the first working laser—that was done by Theodore Maiman of Hughes Research Labs, in 1960.

112 LASERS RELY CRUCIALLY To focus on the essentials here, I'm omitting several details from the analogy. For example, instead of a step stool, there should really be a ladder next to each watermelon, with different rungs for each of the possible excited energy levels that an atom can have. But in many lasers, the more highly excited atoms drop down rapidly and accumulate in the lowest of these rungs; that's what the step stool really represents. Also, photons are not as featureless as seeds or bullets; they have a specific color (corresponding to the wavelength of the light they're carrying), and they behave like waves in many respects. You could think of a photon as having a corrugated appearance, complete with crests and troughs like the ripples on a pond. Furthermore, I'm skipping over the facts that a photon has to have the right color to excite an atom up to a higher rung, or to cause stimulated emission; the laser cavity has to be adjusted to the right length to resonate with the desired wavelength of laser light; the photons ejected by stimulated emission also have the same polarization as the ones that spawned them; and so on.

113 POWER GRID A good introduction, with emphasis on the possible effects of deregulation, is given in Thomas J. Overbye, "Reengineering the electric grid," *American Scientist* 88 (May/June 2000), pp. 220–229. For technical background, see Arthur R. Bergen, *Power Systems Analysis* (Englewood Cliffs, New Jersey: Prentice Hall, 1986).

116 DURING RUSH HOUR For a riveting account of the 1965 Northeast Blackout, see Theodore H. White, "What went wrong? Something called 345 KV," *Life Magazine* 59 (November 19, 1965).

117 COMPUTER CHIP For more about the limitations of synchronous clocking, and the challenges of designing asynchronous chips in which each local circuit runs as fast as it can, see Ivan E. Sutherland and Jo Ebergen, "Computers without clocks," *Scientific American* 287 (August 2002), pp. 62–69; John Markoff, "Computing pioneer challenges the clock," *New York Times* (March 5, 2001).

118 ATOMIC CLOCKS An introduction is available on-line at http://www.boulder.nist. gov/timefreq/index.html, and a more technical discussion is in James C. Bergquist, Steven R. Jefferts, and David J. Wineland, "Time measurement at the millennium," *Physics Today* (March 2001), pp. 37–42.

118 GLOBAL POSITIONING SYSTEM T. A. Herring, "The global positioning system," *Scientific American* 274 (February 1996), pp. 44–50; Anonymous, "Accuracy is addictive," *The Economist* (Technology Quarterly) (March 16, 2002), pp. 24–25.

119 THE PLANETS ARE LOCKED IN ORBITAL RESONANCE Sharon Begley, " 'N sync and a whopper," *Newsweek* (January 22, 2001), pp. 52–53; R. Cowen, "Astronomers find two planetary systems," *Science News* 159 (January 13, 2001), p. 22.

122 EXTINCTION OF THE DINOSAURS A very readable account of the impact theory, and the eventual discovery of the Yucatán crater, is given in Walter Alvarez, *T. Rex and the Crater of Doom* (Princeton: Princeton University Press, 1997).

123 THE ASTEROID BELT Ron Cowen, "A rocky bicentennial: Asteroids come of age," *Science News* 160 (July 28, 2001), pp. 61–63.

124 KIRKWOOD GAPS The role of chaos in creating the gaps was first elucidated in J. Wisdom, "Meteorites may follow a chaotic route to Earth," *Nature* 315 (1985), pp. 731–733.

124 RESONANCE For a recent review, see N. Murray and M. Holman, "The role of chaotic resonances in the solar system," *Nature* 410 (2001), pp. 773–779.

125 WHERE EARTH'S WATER CAME FROM A. Morbidelli et al., "Source regions and timescales for the delivery of water to the Earth," *Meteoritics and Planetary Science* 35 (2000), pp. 1309–1320. For a popular account of the latest thinking about the origin of Earth's water, see Ben Harder, "Water for the rock: Did Earth's oceans come from the heavens?" *Science News* 161 (March 23, 2002), pp. 184–186.

CHAPTER 5 QUANTUM CHORUSES

127 ELECTRICITY My treatment of much of the material in this chapter, from basic electronics to superconductivity, has been heavily influenced by Richard Turton's

engaging exposition in *The Quantum Dot: A Journey into the Future of Microelectronics* (Oxford, England: Oxford University Press, 1995).

128 KAMERLINGH-ONNES R. D. Ouboter, "Heike Kamerlingh-Onnes's discovery of superconductivity," *Scientific American* 276 (March 1997), pp. 98–103.

129 QUANTUM MECHANICS Although physicists have been trying for 70 years to explain the basics of quantum mechanics to a lay audience, no one has ever done a better job than Brian Greene in his best-selling book *The Elegant Universe: Superstrings, Hidden Dimensions, and the Quest for the Ultimate Theory* (New York: W.W. Norton and Company, 1999). His explanations are creative, scientifically honest, and wonderfully pedagogical.

130 RULES OF QUANTUM GROUP BEHAVIOR Richard P. Feynman, Robert B. Leighton, and Matthew Sands, *The Feynman Lectures on Physics, Volume III: Quantum Mechanics* (Reading, Massachusetts: Addison-Wesley, 1965). See Chapter 4 for a discussion of fermions, bosons, the Pauli exclusion principle, why bosons like to crowd together, and a derivation of Planck's blackbody radiation formula from Bose statistics, all done at a level accessible to a strong undergraduate majoring in physics.

131 ALBERT EINSTEIN One of the best scientific biographies of Einstein is Abraham Pais, *Subtle Is the Lord: The Science and the Life of Albert Einstein* (Oxford, England: Oxford University Press, 1982). Einstein's work on what is now called Bose-Einstein condensation is discussed at a technical level in Chapter 23. For more on his correspondence with Bose, see William Blanpied, "Einstein as guru? The case of Bose," in *Einstein: The First Hundred Years,* edited by Maurice Goldsmith, Alan Mackay, and James Woudhuysen (Oxford, England: Pergamon Press, 1980).

131 BOSE HAD ASSUMED NEW RULES FOR COUNTING For a clear explanation of Bose's way of counting all the different configurations of indistinguishable particles, see http://home.achilles.net/~jtalbot/history/einstein.html. The analogy with Peter and Paul is not quite on point, though it suggests that there may be different but equally reasonable ways to count. The real issue that Bose faced was, Given a fixed total energy, how many ways are there of assigning particles to energy levels so that the sum of all the energies is equal to the given total? A nice graphical illustration is shown at http://hyperphysics.phy-astr.gsu.edu/hbase/quantum/disbex.html.

134 "THE THEORY IS PRETTY" Quoted in Pais (1982), p. 432.

134 LESS THAN A MILLIONTH OF A DEGREE The technical feat was not only to reach such temperatures, but also to keep the gas from liquefying or crystallizing before it could condense into the new, exotic state of matter. This required that the gas be extremely dilute, so that its atoms could barely interact.

134 BOSE-EINSTEIN CONDENSATE Eric A. Cornell and Carl E. Wieman, "The Bose-Einstein condensate," *Scientific American* 278 (March 1998), pp. 40–45; Wolfgang

Ketterle, "Experimental studies of Bose-Einstein condensation," *Physics Today* 52 (December 1999), pp. 30–35. An excellent Web site, structured in an entertaining question-and-answer format, is http://www.colorado.edu/physics/2000/bec/.

134 PRESS RELEASE The press release from the Royal Swedish Academy of Sciences is on-line at http://www.nobel.se/physics/laureates/2001/press.html.

134 "OVERLAPPING STEW" George Johnson, "Quantum stew: How physicists are redefining reality's rules," *New York Times* (October 16, 2001), p. F4.

135 HOW SUPERCONDUCTIVITY WORKS The classic paper is J. Bardeen, L. N. Cooper, and J. R. Schrieffer, "Theory of superconductivity," *Physical Review* 108 (1957), pp. 1175–1204. A relatively accessible text is Michael Tinkham, *Introduction to Superconductivity*, 2nd edition (New York: McGraw Hill, 1995).

137 PROPORTIONAL TO $N + 1$ Feynman et al. (1965), vol. III, Section 4.3, explains this rule in a way that makes it look easy, as only Feynman could.

138 HIGH-TEMPERATURE SUPERCONDUCTIVITY J. R. Kirtley and C. C. Tsuei, "Probing high-temperature superconductivity," *Scientific American* 275 (August 1996), pp. 68–73.

139 PRACTICAL APPLICATIONS OF SUPERCONDUCTIVITY Peter Weiss, "Little big wire: High-temperature superconductivity makes a bid for the power grid," *Science News* 158 (November 18, 2000), pp. 330–332; B. Schechter, "Engineering superconductivity. No Resistance: High-temperature superconductors start finding real-world uses," *Scientific American* 283 (August 2000), pp. 32–33; Steven Ashley, "Superconductors heat up," *Mechanical Engineering* (June 1996), pp. 58–63.

140 HE WAS FINDING HIMSELF FASCINATED Josephson's reminiscences are in his acceptance speech for the Nobel Prize, reprinted in B. D. Josephson, "The discovery of tunneling supercurrents," *Science* 184 (1974), pp. 527–530.

140 "A DISCONCERTING EXPERIENCE" Anderson recalls what it was like to teach Josephson in Philip W. Anderson, "How Josephson discovered his effect," *Physics Today* 23 (November 1970), pp. 23–28.

141 JOSEPHSON'S PREDICTION B. D. Josephson, "Possible new effects in superconductive tunneling," *Physics Letters* 1 (1962), pp. 251–253.

143 EQUALLY UNNERVING To appreciate just how amazing this prediction seemed at the time, it helps to hear from one of the protagonists. Anderson (1970) admits that he, Josephson, and Pippard were all "very much puzzled by the meaning of the fact that the current depends on the phase . . . I think it was residual uneasiness on this score that caused the two Brians (Pippard and Josephson) to decide to send the paper to *Physics Letters,* which was just then starting publications, rather than to *Physical Review Letters.*" What he means is that they were all so unsure of Josephson's predictions that they didn't want to send them to the leading journal, in case they turned out to be wrong.

145   A FACE-TO-FACE SHOWDOWN   My account is based entirely on Donald G. McDonald, "The Nobel laureate versus the graduate student," *Physics Today* 54 (2001), pp. 46–51. Giaever's quote appears on p. 49.

146   THEIR MEASUREMENTS   P. W. Anderson and J. M. Rowell, "Probable observation of Josephson superconducting tunneling effect," *Physical Review Letters* 10 (1963), p. 230.

146   FEYNMAN'S ARGUMENT   Richard P. Feynman, Robert B. Leighton, and Matthew Sands, *The Feynman Lectures on Physics, Volume III: Quantum Mechanics* (Reading, Massachusetts: Addison-Wesley, 1965). See Chapter 21 for a discussion of superconductivity, and especially Section 21.9 for an elementary derivation of the Josephson effects.

146   SUPERFLUID HELIUM   N. David Mermin and David M. Lee, "Superfluid helium 3," *Scientific American* 235 (December 1976), pp. 56–71.

147   PHYSICISTS AT THE UNIVERSITY OF CALIFORNIA   S. V. Pereverzev et al., "Quantum oscillations between two weakly coupled reservoirs of superfluid He-3," *Nature* 388 (1997), pp. 449–451. For popular accounts of this work, see P. McClintock, "Quantum mechanics: Whistles from superfluid helium," *Nature* 388 (1997), p. 421, and Michael Brooks, "Liquid genius," *New Scientist* 159 (September 5, 1998), pp. 24–28.

148   DEVICE CALLED A SQUID   John Clarke, "SQUIDs," *Scientific American* 271 (August 1994), pp. 46–53.

149   NEW GENERATION OF SUPERCOMPUTERS   The principles behind Josephson computers are explained in Turton (1995).

150   IBM FAMOUSLY INVESTED   Arthur L. Robinson, "IBM drops superconducting computer project," *Science* 222 (1983), pp. 492–494.

150   THE DREAM OF A JOSEPHSON COMPUTER   S. Hasuo, "Toward the realization of a Josephson computer," *Science* 255 (1992), pp. 301–305.

150   PREOCCUPIED WITH PARANORMAL PHENOMENA   For an interview in which Josephson discusses his interests outside of mainstream physics, see John Gliedman, "The Josephson junction," *Omni* 4 (July 1982), pp. 86–8¢. The spoon bending quote is on p. 116. For more recent snapshots, see John Horgan, "Josephson's inner junction," *Scientific American* 272 (May 1995), pp. 40–41, and http://www.tcm.phy.cam.ac.uk/~bdj10/mm/articles/PWprofile.html. For balance, you should also read Josephson in his own words. His Web page contains a great deal of information about his current views; see http://www.tcm.phy.cam.ac.uk/~bdj10/.

150   HOMEOPATHY   Homeopathy is a system of alternative medicine in which diseases are treated with highly diluted substances that would, if applied in larger doses, cause the same symptoms as the disease itself. Its proponents believe that the remedy

becomes more effective the more diluted it is. Taken to an absurd extreme, a super-dilute solution might not even contain a single molecule of the supposedly active substance—it could be pure water—and yet the homeopathic believers maintain that the potion can still be effective, thanks to a "memory" that the substance imparts on the water molecules. Josephson has supported a scientist named Jacques Benveniste, who claims that this memory of water might have an electromagnetic signature, and that this signature could could be captured electronically, digitized, and then trans-mitted by E-mail, to convert a faraway jar of ordinary water into a homeopathic solution with the desired medicinal properties. Josephson proposed an experiment to test the idea, much to the delight of the physicist Robert Park, a longtime skeptic who had made fun of Benveniste, and who promptly accepted the challenge; see Leon Jaroff, "Homeopathic e-mail," *Time* (May 17, 1999), p. 77. But as of this writ-ing, the experiment still hasn't taken place. I'm not sure why not. The most generous interpretation is that the two sides haven't been able to agree on the protocol. James Randi, a.k.a. "The Amazing Randi," the noted magician, skeptic, and debunker, takes a dimmer view. He accuses Josephson and his homeopathic associates of stalling and finally backing out of the experiment. See Randi's Web page http://www.randi.org/jr/01-26-2001.html, and search his Web site for "Josephson."

151   A SPECIAL SET OF STAMPS   Erica Klarreich, "Stamp booklet has physicists licked," *Nature* 413 (2001), p. 339; Robin McKie, "Royal Mail's Nobel guru in telepathy row," *The Observer* (September 30, 2001). For a spirited and funny counterattack on Josephson's critics by a fine physicist and science writer, see Robert Matthews, "Time travel," *Sunday Telegraph* (London) (November 4, 2001).

CHAPTER 6 BRIDGES

153   RIDING HIS UNICYCLE   This wacky image of Wiener comes from Murray Gell-Mann's recollections of his days as a student at MIT. See George Johnson, *Strange Beauty: Murray Gell-Mann and the Revolution in Twentieth-Century Physics* (New York: Vintage Books, 2000), p. 69.

153   THE FIRST BRIDGE   D. E. McCumber, "Effect of ac impedance on dc voltage-current characteristics of superconductor weak-link junctions," *Journal of Applied Physics* 39 (1968), pp. 3113–3118; W. C. Stewart, "Current-voltage characteristics of Josephson junctions," *Applied Physics Letters* 12 (1968), pp. 277–280.

155   EQUATIONS FOR THE PENDULUM ARE NONLINEAR   The mechanical analog of a Josephson junction is a damped pendulum driven by a constant torque. For a deriva-tion of this analogy, and an analysis of the nonlinear dynamics of both systems, see Sections 4.6 and 8.5 in Steven H. Strogatz, *Nonlinear Dynamics and Chaos: With*

Applications to Physics, Biology, Chemistry, and Engineering (Cambridge, Massachusetts: Perseus Books, 1994).

157 A PARADIGM OF CHAOS B. A. Huberman and J. P. Crutchfield, "Chaotic states of anharmonic systems in periodic fields," *Physical Review Letters* 43 (1979), pp. 1743–1747; D. D'Humieres, M. R. Beasley, B. A. Huberman, and A. Libchaber, "Chaotic states and routes to chaos in the forced pendulum," *Physical Review A* 26 (1982), pp. 3483–3496; N. F. Pedersen and A. Davidson, "Chaos and noise rise in Josephson junctions," *Applied Physics Letters* 39 (1981), pp. 830–832; R. L. Kautz and R. Monaco, "Survey of chaos in the RF-biased Josephson junction," *Journal of Applied Physics* 57 (1985), pp. 875–889.

158 JOSEPHSON ARRAYS For recent reviews, see R. S. Newrock et al., "The two-dimensional physics of Josephson junction arrays," *Solid State Physics: Advances in Research and Applications* 54 (2000), pp. 263–512; C. A. Hamilton, C. J. Burroughs, and S. P. Benz, "Josephson voltage standard: A review," *IEEE Transactions on Applied Superconductivity* 7 (1997), pp. 3756–3761.

159 "SELF-ORGANIZED CRITICALITY" The original paper was Per Bak, Chao Tang, and Kurt Wiesenfeld, "Self-organized criticality: An explanation of 1/f noise," *Physical Review Letters* 59 (1987), pp. 381–384.

159 "SELF-AGGRANDIZING TRIVIALITY" I don't know who came up with this phrase, but I heard it in a lecture given by the physicist Predrag Cvitanovic.

160 STABILITY CHARACTERISTICS OF THE SYNCHRONIZED STATE Peter Hadley, Malcolm R. Beasley, and Kurt Wiesenfeld, "Phase locking of Josephson-junction series arrays," *Physical Review B* 38 (1988), pp. 8712–8719.

161 THE NUMBER GROWS EXTREMELY RAPIDLY Kurt Wiesenfeld and Peter Hadley, "Attractor crowding in oscillator arrays," *Physical Review Letters* 62 (1989), pp. 1335–1338.

163 IT COMES STRAIGHT FROM THE CIRCUIT EQUATIONS The circuit equations are derived and analyzed in K. Y. Tsang, R. E. Mirollo, S. H. Strogatz, and K. Wiesenfeld, "Dynamics of a globally coupled oscillator array," *Physica D* 48 (1991), pp. 102–112. On the last page of the paper, we describe our observations of the unexpected "Russian doll" structure (technically known as a foliation of phase space by nested two-dimensional tori).

168 KURT AND HIS STUDENT S. Nichols and K. Wiesenfeld, "Ubiquitous neutral stability of splay-phase states," *Physical Review A* 45 (1992), pp. 8430–8435.

168 JIM SWIFT J. W. Swift, S. H. Strogatz, and K. Wiesenfeld, "Averaging of globally coupled oscillators," *Physica D* 55 (1992), pp. 239–250.

168 SHINYA WATANABE S. Watanabe and S. H. Strogatz, "Integrability of a globally coupled oscillator array," *Physical Review Letters* 70 (1993), pp. 2391–2394; "Con-

stants of motions for superconducting Josephson arrays," *Physica D* 74 (1994), pp. 197–253.

169   THERE, STARING US IN THE FACE   Kurt Wiesenfeld, Pere Colet, and Steven H. Strogatz, "Synchronization transitions in a disordered Josephson series array," *Physical Review Letters* 76 (1996), pp. 404–407; "Frequency locking in Josephson arrays: Connection with the Kuramoto model," *Physical Review E* 57 (1998), pp. 1563–1569. For a popular account of this work, see Ivars Peterson, "Keeping the beat," *Science News* 149 (April 13, 1996), pp. 236–237.

170   COUPLED LASERS   G. Kozyreff, A. G. Vladimirov, and P. Mandel, "Global coupling with time delay in an array of semiconductor lasers," *Physical Review Letters* 85 (2000), pp. 3809–3812.

170   NEUTRINOS   J. Pantaleone, "Stability of incoherence in an isotropic gas of oscillating neutrinos," *Physical Review D* 58 (1998), article number 073002.

171   MILLENNIUM BRIDGE   I. Sample, "Bad vibrations: How could the designers of a revolutionary bridge miss something so obvious?" *New Scientist* 167 (July 8, 2000), p. 14; Deyan Sudjic, "At last: a bridge you can cross. After a shaky start, the Millennium Bridge is undergoing major surgery. Here, its creators reveal what went wrong and why the blade of light won't wobble when it reopens," *The Observer* (March 11, 2001).

171   "A BLADE OF LIGHT"   Lord Foster is quoted in Matthew Jones, "Survey: The South Bank reborn: Brave vision of blade of light," *Financial Times (London)* (May 9, 2000), p. 2.

172   ARUP, THE ENGINEERING FIRM   Arup's explanation of what caused the bridge to wobble is given at http://www.arup.com/MillenniumBridge/. A simulation of the bridge's motion is available on-line at http://www2.eng.cam.ac.uk/~gm249/MillenniumBridge/.

175   LETTER TO THE EDITOR   Brian Josephson, "Out of step on the bridge," *The Guardian (London)* (June 14, 2000), Guardian Leader Pages, p. 23.

CHAPTER 7 SYNCHRONIZED CHAOS

179   "THAT LITTLE MODEL"   E. N. Lorenz, "Deterministic nonperiodic flow," *Journal of the Atmospheric Sciences* 20 (1963), pp. 130–141.

179   THE MODERN FIELD OF CHAOS THEORY   The best introduction to chaos theory is still James Gleick's captivating classic, *Chaos: Making a New Science* (New York: Viking, 1987). It's full of wonderful inside stories about scientists at work, and Gleick's explanations are both accessible and accurate. Lorenz's own view of the subject is given in Edward N. Lorenz, *The Essence of Chaos* (Seattle: University of Washington Press, 1993). For those seeking an elementary introduction to the

mathematics and science of chaos, see Steven H. Strogatz, *Nonlinear Dynamics and Chaos: With Applications to Physics, Biology, Chemistry, and Engineering* (Cambridge, Massachusetts: Perseus Books, 1994). The Lorenz equations are discussed in Chapter 9.

180   HYPERION   J. Wisdom, S. J. Peale, and F. Mignard, "The chaotic rotation of Hyperion," *Icarus* 58 (1984), pp. 137–152.

183   BUTTERFLY EFFECT   E. N. Lorenz, "Predictability: Does the flap of a butterfly's wings in Brazil set off a tornado in Texas?" Address at the annual meeting of the American Association for the Advancement of Science in Washington, December 29, 1979.

184   CHAOS PROMISED TO BE USEFUL   W. L. Ditto and L. M. Pecora, "Mastering chaos," *Scientific American* 269 (August 1993), pp. 78–84.

185   DESKTOP WATERWHEEL   Strogatz (1994), Section 9.1.

188   "FOR WANT OF A NAIL"   Cited in this context by Gleick (1987), p. 23, who in turn cites an article by Norbert Wiener.

190   THE LYAPUNOV TIME   The conceptual importance of the Lyapunov time is discussed in J. Lighthill, "The recently recognized failure of predictability in Newtonian dynamics," *Proceedings of the Royal Society of London, Series A: Mathematical, Physical, and Engineering Sciences* 407 (1986), pp. 35–50.

190   SOLAR SYSTEM   Its Lyapunov time is estimated in G. Sussman and J. Wisdom, "Chaotic evolution of the solar system," *Science* 257 (1992), pp. 56–62.

191   "STRANGE ATTRACTOR"   For a clear introduction to strange attractors, see J. P. Crutchfield, J. D. Farmer, N. H. Packard, and R. S. Shaw, "Chaos," *Scientific American* 255 (December 1986), pp. 46–&c.

194   WHEN LOU PECORA BEGAN TO DAYDREAM   Pecora told me the colorful story of his work on synchronized chaos (with Tom Carroll) during two phone interviews conducted on January 27 and February 1, 2002.

197   HIS SCHEME   The seminal paper on synchronized chaos is L. M. Pecora and T. L. Carroll, "Synchronization in chaotic systems," *Physical Review Letters* 64 (1990), pp. 821–824. For a review of more recent work, see L. M. Pecora et al., "Fundamentals of synchronization in chaotic systems: Concepts and applications," *Chaos* 7 (1997), pp. 520–543. As with many significant discoveries, we now know that Pecora and Carroll were not actually the first to notice the possibility of synchronized chaos. See, for example, H. Fujisaka and T. Yamada, "Stability theory of synchronized motion in coupled-oscillator systems," *Progress of Theoretical Physics* 69 (1983), pp. 32–47, and V. S. Afraimovich, N. N. Verichev, and M. I. Rabinovich, "General synchronization," *Radiophysics and Quantum Electronics* 29 (1986), pp. 795–803. But those contributions went largely unnoticed, perhaps because they did not emphasize the novelty of the phenomenon or its potential importance for communications.

201   CUOMO AND OPPENHEIM'S PAPER   Kevin M. Cuomo and Alan V. Oppenheim, "Circuit implementation of synchronized chaos with applications to communications," *Physical Review Letters* 71 (1993), pp. 65–68; K. M. Cuomo, A. V. Oppenheim, and S. H. Strogatz, "Synchronization of Lorenz-based chaotic circuits with applications to communications," *IEEE Transactions on Circuits and Systems II: Analog and Digital Signal Processing* 40 (1993), pp. 626–633. A popular account of the use of chaos for private communications appeared in J.C.G. Lesurf, "Electronics: Chaos in harness," *Nature* 365 (1993), pp. 604–605.

203   FOR PEOPLE USING CELLULAR PHONES   Steve Boggan, "Bugging: Can you hear me? Yes, darling, and so can an awful lot of other people," *The Independent* (London) (January 17, 1993); Susan Levine, "Eavesdropping on cellular calls is illegal but easy," *The Washington Post* (January 11, 1997), p. A01; Juliet Eilperin, "Hill tape dispute allowed to continue," *The Washington Post* (January 9, 2002), p. A17.

204   KEVIN SHORT   Kevin M. Short, "Steps toward unmasking secure communications," *International Journal of Bifurcation and Chaos* 4 (1994), pp. 959–977; J. B. Geddes, K. M. Short, and K. Black, "Extraction of signals from chaotic laser data," *Physical Review Letters* 83 (1999), pp. 5389–5392.

204   CHAOTIC COMMUNICATIONS USING LASERS   G. D. VanWiggeren and R. Roy, "Communication with chaotic lasers," *Science* 279 (1998), pp. 1198–1200. For a commentary on this article, see D. J. Gauthier, "Chaos has come again," *Science* 279 (1998), pp. 1156–1157.

CHAPTER 8 SYNC IN THREE DIMENSIONS

206   *THE GEOMETRY OF BIOLOGICAL TIME*   Arthur T. Winfree, *The Geometry of Biological Time* (New York: Springer-Verlag, 1980).

207   DATA FROM HIS OWN MOTHER   Shown on p. 453 of Winfree (1980), in a section titled "Statistics ('Am I Overdue?!')." Winfree once told me that his mother Dorothy kept accurate records of all her menstrual periods because she was a practicing Catholic who used the rhythm method of birth control.

207   "NIXON CHOSE THAT WEEK TO INVADE CAMBODIA"   Winfree (1980), p. 291.

210   THE INTESTINE   Winfree (1980), pp. 325–329, contains a discussion of neuromuscular wave propagation in the small intestine, regarded as a one-dimensional continuum of oscillators.

210   THE STOMACH   The literature supporting the view that the stomach is a two-dimensional bag of oscillators is discussed on pp. 329–330 of Winfree (1980).

210   THE HEART   For Winfree's views on three-dimensional waves in the heart, see A. T. Winfree, *When Time Breaks Down: The Three-Dimensional Dynamics of Electro-*

chemical Waves and Cardiac Arrhythmias (Princeton, New Jersey: Princeton University Press, 1987).

210   CARDIOLOGISTS HAD KNOWN FOR DECADES   G. R. Mines, "On circulating excitations on heart muscles and their possible relation to tachycardia and fibrillation," *Transactions of the Royal Society of Canada* 4 (1914), pp. 43–53; W. E. Garrey, "Nature of fibrillary contraction in the heart," *American Journal of Physiology* 33 (1914), pp. 397–414.

210   "CIRCUS MOVEMENTS"   M. A. Allessie, F. I. M. Bonke, and F. J. Schopman, "Circus movement in rabbit atrial muscle as a mechanism of tachycardia," *Circulation Research* 33 (1973), pp. 54–62.

211   DIE SUDDENLY   A. T. Winfree, "Sudden cardiac death: A problem in topology," *Scientific American* 248 (May 1983), pp. 144–&; M. S. Eisenberg, L. Bergner, A. P. Hallstrom, and R. O. Cummins, "Sudden cardiac death," *Scientific American* 254 (May 1986), pp. 37–&.

212   ZHABOTINSKY SOUP   A. N. Zaikin and A. M. Zhabotinsky, "Concentration wave propagation in two-dimensional liquid-phase self-oscillating systems," *Nature* 225 (1970), pp. 535–537.

214   A TALE OF DOGMA, DISAPPOINTMENT, AND ULTIMATE VINDICATION   A. T. Winfree, "The prehistory of the Belousov-Zhabotinsky oscillator," *Journal of Chemical Education* 61 (1984), pp. 661–663.

216   SPIRAL WAVES   A. T. Winfree, "Spiral waves of chemical activity," *Science* 175 (1972), pp. 634–&; "Rotating chemical reactions," *Scientific American* 230 (June 1974), pp. 82–&.

217   JELLYFISH   A. G. Mayer, "Rhythmical pulsation in scyphomedusae," *Papers of the Tortugas Laboratory of the Carnegie Institution of Washington* 1 (1908), pp. 115–131.

219   A NEW KIND OF SPIRAL WAVE   K. I. Agladze and V. I. Krinsky, "Multi-armed vortices in an active chemical medium," *Nature* 296 (1982), pp. 424–426.

219   SCROLL WAVE   A. T. Winfree, "Scroll-shaped waves of chemical activity in three dimensions," *Science* 181 (1973), pp. 937–939. The first direct visualization of a scroll ring appeared in B. J. Welsh, J. Gomatam, and A. E. Burgess, "Three-dimensional chemical waves in the Belousov-Zhabotinskii reaction," *Nature* 304 (1983), pp. 611–614.

220   WINFREE WONDERED   Winfree (1980), pp. 254–257.

221   NONSENSE PICTURE IN THE STYLE OF ESCHER   For the nonsense picture, along with accurate pictures of scroll rings, see S. H. Strogatz, M. L. Prueitt, and A. T. Winfree, "Exotic shapes in chemistry and biology," *IEEE Computer Graphics and Applications* 4 (1984), pp. 66–69.

223   KNOTS WERE HARD   A. T. Winfree and S. H. Strogatz, "Singular filaments organize

chemical waves in three dimensions. III. Knotted waves," *Physica D* 9 (1983), pp. 333–345.

225   LINKING NUMBER   For a review of the mathematics needed to understand the structure of scroll waves, see J. J. Tyson and S. H. Strogatz, "The differential geometry of scroll waves," *International Journal of Bifurcation and Chaos* 1 (1991), pp. 723–744.

226   THE EXCLUSION PRINCIPLE   A. T. Winfree and S. H. Strogatz, "Singular filaments organize chemical waves in three dimensions. IV. Wave taxonomy." *Physica D* 13 (1984), pp. 221–233; "Organizing centers for three-dimensional chemical waves," *Nature* 311 (1984), pp. 611–615. More elegant proofs of the exclusion principle were later found; see A. T. Winfree, E. M. Winfree, and H. Seifert, "Organizing centers in a cellular excitable medium," *Physica D* 17 (1985), pp. 109–115.

226   MEANDER   The meandering of spiral waves is discussed in L. Ge et al., "Transition from simple rotating chemical spirals to meandering and traveling spirals," *Physical Review Letters* 77 (1996), pp. 2105–2108, and in M. Woltering, R. Girnus, and M. Markus, "Quantification of turbulence in the Belousov-Zhabotinsky reaction by monitoring wave tips," *Journal of Physical Chemistry A* 103 (1999), pp. 4034–4037. A key theoretical contribution was made by D. Barkley, "Euclidean symmetry and the dynamics of rotating spiral waves," *Physical Review Letters* 72 (1994), pp. 164–167.

226   THE HOLY GRAIL REMAINS CARDIAC ARRHYTHMIAS   For a sample of recent thinking, see the Special Focus issue of *Chaos,* March 1998. Also, see A. T. Winfree, "Electrical turbulence in three-dimensional heart muscle," *Science* 266 (1994), pp. 1003–1006; A. Garfinkel et al., "Quasiperiodicity and chaos in cardiac fibrillation," *Journal of Clinical Investigation* 99 (1997), pp. 305–314; F. X. Witkowski et al., "Spatiotemporal evolution of ventricular fibrillation," *Nature* 392 (1998), pp. 78–82; A. Panfilov and A. Pertsov, "Ventricular fibrillation: Evolution of the multiple-wavelet hypothesis," *Philosophical Transactions of the Royal Society of London, Series A: Mathematical, Physical, and Engineering Sciences* 359 (2001), pp. 1315–1325; V. N. Biktashev et al., "Three-dimensional organisation of re-entrant propagation during experimental ventricular fibrillation," *Chaos, Solitons, and Fractals* 13 (2002), pp. 1713–1733.

227   HOW LINKED AND KNOTTED SCROLL WAVES WOULD MOVE   A. T. Winfree, "Persistent tangles of vortex rings in excitable media," *Physica D* 84 (1995), pp. 126–147; J. P. Keener and J. J. Tyson, "The dynamics of scroll waves in excitable media," *SIAM Review* 34 (1992), pp. 1–39; D. Margerit and D. Barkley, "Selection of twisted scroll waves in three-dimensional excitable media," *Physical Review Letters* 86 (2001), pp. 175–178. An extensive review of scroll-wave dynamics appears in the updated version of *The Geometry of Biological Time* (2nd edition, 2001).

227   THE BASIC LOCALIZED SOLUTIONS   A. T. Winfree, "Stable particle-like solutions to the nonlinear wave equations of 3-dimensional excitable media," *SIAM Review* 32 (1990), pp. 1–53. An intriguing and closely related study from wave physics is M. V. Berry and M. R. Dennis, "Knotted and linked phase singularities in monochromatic waves," *Proceedings of the Royal Society of London, Series A: Mathematical, Physical, and Engineering Sciences* 457 (2001), pp. 2251–2263.

227   OPTICAL TOMOGRAPHY   A. T. Winfree et al., "Quantitative optical tomography of chemical waves and their organizing centers," *Chaos* 6 (1996), pp. 617–626. For another promising approach, see A. L. Cross et al., "Three dimensional imaging of the Belousov-Zhabotinsky reaction using magnetic resonance," *Magnetic Resonance Imaging* 15 (1997), pp. 719–725.

CHAPTER 9 SMALL-WORLD NETWORKS

229   JOHN GUARE'S 1990 PLAY   John Guare, *Six Degrees of Separation* (New York: Vintage Books, 1990).

229   THREE INEBRIATED FRATERNITY BROTHERS   Ann Oldenburg, "A thousand links to Kevin Bacon: Game calculates actor's connection," *USA Today* (October 18, 1996), p. 5D; Mel Gussow, "Are actors all related? Or is it just Kevin Bacon?" *New York Times* (September 19, 1996), p. C13.

230   MARLON BRANDO   Anonymous, "Media: Six degrees from Hollywood," *Newsweek* (October 11, 1999), p. 6.

230   "SIX DEGREES OF MONICA"   David Kirby and Paul Sahre, "Six degrees of Monica," *New York Times* (February 21, 1998), p. A11.

230   BLACKOUTS IN 11 STATES   Western Systems Coordinating Council (WSCC), "Disturbance report for the power system outage that occurred on the Western Interconnection on August 10th, 1996 at 1548 PAST," (October 1996). Available at http://www.wscc.com.

231   FOCUS OF MOLECULAR BIOLOGY   Two thoughtful papers about the coming era of genetic and biochemical networks are L. H. Hartwell, J. J. Hopfield, S. Leibler, and A. W. Murray, "From molecular to modular cell biology," *Nature* 402 (1999), pp. C47–52, and U. S. Bhalla and R. Iyengar, "Emergent properties of networks of biological signalling pathways," *Science* 283 (1999), pp. 381–387. For a sense of how befuddling these networks are going to be, see K. W. Kohn, "Molecular interaction map of the mammalian cell cycle control and DNA repair systems," *Molecular Biology of the Cell* 10 (1999), pp. 2703–2734.

232   THE FIRST COMPARATIVE STUDY   Duncan J. Watts and Steven H. Strogatz, "Collective dynamics of 'small-world' networks," *Nature* 393 (1998), pp. 440–442. A fuller

presentation is given in Duncan J. Watts, *Small Worlds: The Dynamics of Networks Between Order and Randomness* (Princeton, New Jersey: Princeton University Press, 1999).

232 THE STUDY OF COMPLEX NETWORKS Three recent books survey this emerging field in an entertaining and accessible fashion: Mark Buchanan, *Nexus: Small Worlds and the Groundbreaking Science of Networks* (New York: W.W. Norton & Company, 2002); Albert-László Barabási, *Linked: The New Science of Networks* (Cambridge, Massachusetts: Perseus, 2002); and Duncan J. Watts, *Six Degrees: The Science of a Connected Age* (New York: W.W. Norton & Company, 2003). For an overview aimed at a scientific audience, see Steven H. Strogatz, "Exploring complex networks," *Nature* 410 (2001), pp. 268–276.

234 HOW MALE CRICKETS MANAGE TO CHIRP TOGETHER T. J. Walker, "Acoustic synchrony: Two mechanisms in the snowy tree cricket," *Science* 166 (1969), pp. 891–894. For a detailed study of synchronous chirping in a related species, see E. Sismondo, "Synchronous, alternating, and phase-locked stridulation by a tropical katydid," *Science* 249 (1990), pp. 55–58. The evolutionary significance of synchronous chorusing is discussed by M. D. Greenfield, "Synchronous and alternating choruses in insects and anurans: Common mechanisms and diverse functions," *American Zoologist* 34 (1994), pp. 605–615.

237 IDEALIZED MODEL OF GENE NETWORKS Stuart A. Kauffman, "Metabolic stability and epigenesis in randomly constructed genetic nets," *Journal of Theoretical Biology* 22 (1969), pp. 437–467. For a popular exposition, see Stuart A. Kauffman, *At Home in the Universe: The Search for Laws of Self-Organization and Complexity* (Oxford, England: Oxford University Press, 1995).

243 A TINY WORM Nicholas Wade, "Dainty worm tells secrets of human genetic code," *New York Times* (June 24, 1997). This worm even has its own Web page: http://elegans.swmed.edu/.

244 NERVOUS SYSTEM HAD BEEN COMPLETELY MAPPED J. G. White, E. Southgate, J. N. Thomson, and S. Brenner, "The structure of the nervous system of *Caenorhabditis elegans*," *Proceedings of the Royal Society of London, Series B: Biological Sciences* 314 (1986), pp. 1–340.

244 AVAILABLE ON A FLOPPY DISKETTE The diskette containing the complete map of the worm's nervous system comes with T. B. Achacoso and W. S. Yamamoto, *AY's Neuroanatomy of C. elegans for Computation* (Boca Raton, Florida: CRC Press, 1992).

245 JOHN GUARE HIMSELF Beth Saulnier, "Small world," *Cornell Magazine* 101 (July/August 1998), pp. 24–29. Guare is quoted on p. 26.

245 "SMALL-WORLD PROBLEM" A remarkably prescient formulation was given by Ithiel de Sola Pool, a political scientist at MIT, and Manfred Kochen, a mathematician at

IBM, in their paper "Contacts and influence," *Social Networks* 1 (1978), pp. 1–51. This paper was drafted in 1958, and circulated informally among social scientists for two decades before being published. Milgram himself was inspired by it. Pool and Kochen understood the simple case of a completely random network, and they tried to deal with the complications introduced by clustering, but they couldn't quite make their way through the mathematical maze. For more about the social science literature on this problem, see *The Small World*, edited by Manfred Kochen (Norwood, New Jersey: Ablex, 1989).

245   STANLEY MILGRAM   Stanley Milgram, "The small world problem," *Psychology Today* 2 (1967), pp. 60–67. Sometimes people dismiss social science as nothing more than an academic version of common sense, but the work of Milgram refutes that charge. He was unafraid to ask the big questions, and the results he obtained were anything but obvious. His most famous experiments dealt with obedience to authority. Under the pretense of investigating the effects of punishment on short-term memory, he asked subjects (the "teachers") to administer what they thought were painful electrical shocks to other people (the "learners"), increasing the voltage after each wrong answer to a word-association problem. Of course, no shocks were actually delivered; the learners were actors who were paid to feign agony. The results were profoundly disturbing. Many apparently normal people would shock another person to death, just because a man in a white coat requested it. Most of Milgram's other experiments were not so grim; they typically involved a mix of playfulness and theater—almost a *Candid Camera* approach to social psychology. In one experiment, he sent his graduate students to ride on the New York subway, where they'd ask people to give up their seats without offering any reason. Most New Yorkers were surprisingly compliant; even more unexpected was that the experimenter making the unwarranted request felt tremendous stress. (When Milgram himself tried it, he said "the words seemed lodged in my trachea . . . I could feel my face blanching. I was not role-playing. I actually felt as if I were going to perish.") In another experiment, designed to test the drawing power of crowds of different sizes, he had his confederates stand on the sidewalk and look up at the sixth-floor window of an office building across the street, to see how many other people would join them in gazing off into empty space. For a collection of his essays and articles, all of which are fascinating and eminently readable, see Stanley Milgram, *The Individual in a Social World: Essays and Experiments*, second edition, edited by John Sabini and Maury Silver (New York: McGraw-Hill, 1992).

246   THEY REMAIN INCONCLUSIVE   Judith S. Kleinfeld, "The small world problem," *Society* 39 (2002), pp. 61–66.

246   PAUL ERDŐS   Paul Hoffman, *The Man Who Loved Only Numbers: The Story of Paul Erdős and the Search for Mathematical Truth* (New York: Hyperion, 1998).

246 "ERDŐS NUMBER" Caspar Goffman, "And what is your Erdős number?" *American Mathematical Monthly* 76 (1969), p. 791. A Web site for Erdős numbers, containing lots of amusing trivia and mathematical entertainment, is http://www.oakland.edu/ ~grossman/erdoshp.html.

247 THE ORACLE OF BACON You can play the Kevin Bacon game on-line at http://www.cs.virginia.edu/oracle/. The 1,000 best-connected actors are listed at http://www.cs.virginia.edu/oracle/center_list.html.

248 ONE OF THE SIMPLEST MODELS H. Sakaguchi, S. Shinomoto, and Y. Kuramoto, "Local and global self-entrainments in oscillator lattices," *Progress in Theoretical Physics* 77 (1987), pp. 1005–1010.

249 WE FOUND THAT A TINY PERCENTAGE The results are given in Chapter 9 of Watts (1999).

250 "DENSITY CLASSIFICATION PROBLEM" M. Mitchell, J. P. Crutchfield, and P. T. Hraber, "Evolving cellular automata to perform computations—mechanisms and impediments," *Physics D* 75 (1994), pp. 361–391; James P. Crutchfield and Melanie Mitchell, "The evolution of emergent computation," *Proceedings of the National Academy of Sciences USA* 92 (1995), pp. 10742–10746.

251 SMALL-WORLD NETWORK OF BULBS The results are summarized in Chapter 7 of Watts (1999).

251 INFECTIOUS DISEASES J. Wallinga, K. J. Edmunds, and M. Kretzschmar, "Perspective: Human contact patterns and the spread of airborne infectious diseases," *Trends in Microbiology* 7 (1999), pp. 372–377; M. J. Keeling, "The effects of local spatial structure on epidemiological invasions," *Proceedings of the Royal Society of London, Series B: Biological Sciences* 266 (1999), pp. 859–867; M. Boots and A. Sasaki, " 'Small worlds' and the evolution of virulence: infection occurs locally and at a distance," *Proceedings of the Royal Society of London, Series B: Biological Sciences* 266 (1999), pp. 1933–1938.

252 THE SPREAD OF AIDS Randy Stilts, *And the Band Played On: Politics, People, and the AIDS Epidemic* (New York: St. Martins Press, 1987).

252 FOOT-AND-MOUTH DISEASE Mark Woolhouse and Alex Donaldson, "Managing foot-and-mouth," *Nature* 410 (2001), p. 515.

253 MARK GRANOVETTER Mark S. Granovetter, "The strength of weak ties," *American Journal of Sociology* 78 (1973), pp. 1360–1380. The quote is transcribed from the BBC Radio program "Living by Numbers," broadcast on July 1, 1999.

254 ANATOMY OF THE WORLD WIDE WEB R. Albert, H. Jeong, and A.-L. Barabási, "Diameter of the World Wide Web," *Nature* 401 (1999), pp. 130–131. A much more comprehensive study of the Web's connectivity has now been performed, prompted in part by the work of Barabási and his students; see A. Broder et al., "Graph structure in the Web," *Computer Networks* 33 (2000), pp. 309–320.

255 "POWER LAW" For a lively introduction to power laws in all their guises, see Manfred Schoeder, *Fractals, Chaos, Power Laws: Minutes from an Infinite Paradise* (New York: W. H. Freeman, 1991).

255 THE ORIGIN OF POWER LAWS REMAINS CONTROVERSIAL At least seven different physical mechanisms can generate power laws. In that sense, the experimental observation of a power law is not, in itself, a stringent test of any theory that predicts one. For a refreshingly clear-minded discussion of this point, see Mark Newman, "Applied mathematics: The power of design," *Nature* 405 (2000), pp. 412–413.

255 BARABÁSI AND HIS TEAM OFFERED Albert-László Barabási and Réka Albert, "Emergence of scaling in random networks," *Science* 286 (1999), pp. 509–512.

256 AN EXPLOSION OF EMPIRICAL STUDIES For a review, see Réka Albert and Albert-László Barabási, "Statistical mechanics of complex networks," *Reviews of Modern Physics* 74 (2002), pp. 47–97. This is also an excellent introduction to the mathematical techniques used in the field.

256 A NETWORK FOR THE ENGLISH LANGUAGE R. F. I. Cancho and R. V. Solé, "The small world of human language," *Proceedings of the Royal Society of London, Series B: Biological Sciences* 268 (2001), pp. 2261–2265.

257 A PHYSICIST FRIEND Charlie Marcus, a physics professor at Harvard.

257 CIRCUITS TEND TO BE WIRED IN A SMALL-WORLD FASHION R. F. I. Cancho, C. Janssen, and R. V. Solé, "Topology of technology graphs: Small world patterns in electronic circuits," *Physical Review E* 64 (2001), article number 046119 Part 2.

257 RESISTANT TO RANDOM FAILURES, YET VULNERABLE TO DELIBERATE ATTACK This property of scale-free networks was first pointed out by Réka Albert, Hawoong Jeong, and Albert-László Barabási, "Error and attack tolerance of complex networks," *Nature* 406 (2000), pp. 378–382, on the basis of computer simulations. A rigorous mathematical treatment was developed independently by R. Cohen, K. Erez, D. ben-Avraham, and S. Havlin, "Resilience of the Internet to random breakdowns," *Physical Review Letters* 85 (2000), pp. 4626–4628, and by D. S. Callaway, M.E.J. Newman, S.H. Strogatz, and D.J. Watts, "Network robustness and fragility: Percolation on random graphs," *Physical Review Letters* 85 (2000), pp. 5468–5471.

258 PROTEIN INTERACTIONS IN YEAST H. Jeong, S. P. Mason, A.-L. Barabási, and Z. N. Oltvai, "Lethality and centrality in protein networks," *Nature* 411 (2001), pp. 41–42.

258 AN ARTICLE IN *BUSINESS WEEK* Nellie Andreeva, "Do the math—It is a small world," *Business Week* (August 17, 1998), pp. 54–55.

258 FBI FORENSIC SCIENTIST Max Houck, now the director of West Virginia University's new Forensic Science Initiative.

CHAPTER 10 THE HUMAN SIDE OF SYNC

260   ALAN ALDA   For a fuller statement of what he finds so fascinating about fads, see http://www.edge.org/q2002/q_alda.html.

263   ANATOL RAPOPORT   Anatol Rapoport, "Spread of information through a population with sociostructural bias," *Bulletin of Mathematical Biophysics* 15 (1953), pp. 523–543.

263   "TIPPING POINT"   Although this felicitous phrase was first used by Morton Grodzins, "Metropolitan segregation," *Scientific American* 197 (October 1957), pp. 33–41, the classic paper on the tipping point is generally acknowledged to be Thomas Schelling, "Dynamic models of segregation," *Journal of Mathematical Sociology* 1 (1971), pp. 143–186. Both Schelling and Grodzins sought to explain the abruptness of white flight from racially mixed neighborhoods, once a critical number of black people move in. What's so counterintuitive about this phenomenon is that a seemingly harmless individual preference (a slight desire to have some neighbors like yourself) can snowball into a drastic and undesirable social outcome (total racial segregation). The wider public first became aware of the concept of the tipping point a few years ago, thanks to Malcolm Gladwell's best-selling book *The Tipping Point: How Little Things Can Make a Big Difference* (New York: Little Brown, 2000). Gladwell is a terrific raconteur, and it's fun to follow him as he examines hits, fads, social movements, epidemics, and other phenomena that depend on contagion in one form or another.

263   MARK GRANOVETTER   M. Granovetter, "Threshold models of collective behavior," *American Journal of Sociology* 83 (1978), pp. 1420–1443.

264   DUNCAN WATTS   Duncan J. Watts, "A simple model of global cascades on random networks," *Proceedings of the National Academy of Sciences USA* 99 (2002), pp. 5766–5771.

267   MARKETING LANGUAGE   Everett M. Rogers, *Diffusion of Innovations*, 4th edition (New York: Free Press, 1995).

268   TRAFFIC   For a good summary of the recent work on the self-organizing aspects of traffic patterns, see Peter Weiss, "Stop-and-go science," *Science News* 156 (July 3, 1999), pp. 8–10.

269   A STATE OF CRYSTALLINE HARMONY   Dirk Helbing and Bernardo Huberman, "Coherent moving states in highway traffic," *Nature* 396 (1998), pp. 738–740. For a popular account of this work, see Robert Kunzig, "The physics of traffic: Curing congestion," *Discover* (March 1999), pp. 31–32.

270   DIFFERENT FORM OF SYNCHRONIZED TRAFFIC   B. S. Kerner and H. Rehborn, "Experimental properties of phase transitions in traffic flow," *Physical Review Letters* 79 (1997), pp. 4030–4033.

271 COMPUTER SIMULATIONS LATER DEMONSTRATED   H. Y. Lee, H.-W. Lee, and D. Kim, "Onset of synchronized traffic flow on highways and its dynamic phase transitions," *Physical Review Letters* 81 (1998), pp. 1130–1133.

271 CLAPPING IN UNISON   Z. Néda, E. Ravasz, T. Vicsek, Y. Brechet, and A.-L. Barabási, "The sound of many hands clapping," *Nature* 403 (2000), pp. 849–850; Z. Néda, E. Ravasz, Y. Brechet, T. Vicsek, and A.-L. Barabási, "Physics of the rhythmic applause," *Physical Review E* 61 (2000), pp. 6987–6992. A popular account appeared in Josie Glausiusz, "The mathematics of applause," *Discover* (July 2000), p. 32.

273 MODELS OF ARTIFICIAL SOCIETIES   Jonathan Rauch, "Seeing around corners," *The Atlantic Monthly* (April 2002), pp. 35–48.

274 WHEN THE MOON IS FULL   The lunar effect is thoroughly debunked by I. W. Kelly, James Rotton, and Roger Culver, "The moon was full and nothing happened: A review of studies on the moon and human behavior and human belief," in *The Outer Edge*, edited by J. Nickell, B. Karr and T. Genoni (Amherst, New York: CSICOP, 1996). For an evenhanded summary of the data, see http://faculty.washington. edu/chudler/moon.html, and for more debunking, see http://skepdic.com/fullmoon. html.

275 "BIORHYTHMS"   The quack notion of biorhythm is discussed (and dismissed) by A. T. Winfree, *The Timing of Biological Clocks* (New York: Scientific American Books, 1987), pp. 6–8.

275 "SYNCHRONICITY"   Carl G. Jung, *Synchronicity: An Acausal Connecting Principle*, translated by R. F. C. Hull; Bollingen Series (Princeton, New Jersey: Princeton University Press, 1973).

275 COINCIDENCES   P. Diaconis and F. Mosteller, "Methods for studying coincidences," *Journal of the American Statistical Association* 84 (1989), pp. 853–861. A sensible analysis is also given on-line at http://www.csicop.org/si/9809/coincidence.html.

276 AN OUTRAGEOUS EXPERIMENT   Norbert Wiener, *Nonlinear Problems in Random Theory* (Cambridge, Massachusetts: MIT Press, 1958), pp. 71–72, and *Cybernetics*, p. 198.

276 POKÉMON   Janet Snyder, "Monster TV cartoon illness mystifies Japan," *Reuters* (December 17, 1997). An analysis in the medical literature was given by T. Takahashi and Y. Tsukahara, "Pocket Monster incident and low luminance visual stimuli: Special reference to deep red flicker stimulation," *Acta Paediatrica Japonica* 40 (1998), pp. 631–637. For links to the on-line coverage of this incident, see http://www.virtualpet.com/vp/farm/pmonster/seizures/seizures.html. An excellent general reference about photosensitive epilepsy is http://www.epilepsytoronto.org/ people/eaupdate/vol9-3.html.

277 NEURAL SYNCHRONY For two readable summaries of this controversial field, see Bruce Bower, "All fired up: Perception may dance to the beat of collective neuronal rhythms," *Science News* 153 (February 21, 1998), pp. 120–121, and B. Schechter, "How the brain gets rhythm," *Science* 274 (1996), pp. 339–340.

277 "BINDING PROBLEM" C. von der Malsburg, "The what and why of binding: The modeler's perspective," *Neuron* 24 (1999), pp. 95–104. This entire issue of *Neuron* (September 1999) is devoted to the binding problem. Von der Malsburg's original paper from 1981 is hard to find, but it is reprinted in *Models of Neural Networks II*, edited by E. Domany, J. L. van Hemmen, and K. Schulten (Berlin: Springer-Verlag, 1994).

278 "THE MIND WOULD BE INVISIBLE" Quoted in B. Schechter, "How the brain gets rhythm," *Science* 274 (1996), pp. 339–340.

278 A TEAM OF NEUROSCIENTISTS C. M. Gray, A. K. Engel, P. Konig, and W. Singer, "Oscillatory responses in cat visual cortex exhibit inter-columnar synchronization which reflects global stimulus properties," *Nature* 338 (1989), pp. 334–337.

279 IT WAS HARD TO UNDERSTAND For an introduction to the mathematical puzzles about long-range synchrony in the brain, and hints about how they might be resolved, see Nancy Kopell, "We got rhythm: Dynamical systems of the nervous system," *Notices of the American Mathematical Society* 47 (2000), pp. 6–16, and Barry A. Cipra, "It's got a beat, and you can think to it," *SIAM News* 34 (April 2001), p. 1–&.

280 SHORT-TERM MEMORY Jürgen Fell et al., "Human memory formation is accompanied by rhinal-hippocampal coupling and decoupling," *Nature Neuroscience* 4 (2001), pp. 1259–1264.

281 *MISTOOK HIS WIFE FOR A HAT* Oliver Sacks, *The Man Who Mistook His Wife for a Hat* (New York: Simon & Schuster, 1988).

281 NEUROSCIENTISTS LED BY FRANCISCO VARELA Eugenio Rodriguez et al., "Perception's shadow: Long-distance synchronization of human brain activity," *Nature* 397 (1999), pp. 430–433.

283 CONSCIOUSNESS MAY BE THE SUBJECTIVE EXPERIENCE Walter J. Freeman, "The physiology of perception," *Scientific American* 264 (February 1991), pp. 78–85. This is also a very readable account of Freeman's pioneering work, which provided some of the earliest experimental evidence linking brain rhythms to perception.

283 "THE ZOMBIE WITHIN" Christof Koch and Francis Crick, "The zombie within," *Nature* 411 (2001), p. 893.

INDEX

PENGUIN SCIENCE

SCIENCE: A HISTORY   JOHN GRIBBIN

The enthralling story of the men and women who changed the way we see the world, and the turbulent times they lived in: from Galileo, tried by the Inquisition for his ideas, to Newton, who wrote his rivals out of the history books; from Marie Curie, forced to work apart from male students for fear that she might excite them, to Louis Agassiz, who marched his colleagues up a mountain to prove that ice ages had occurred. Filled with pioneers, visionaries, eccentrics and madmen, this is the history of science as it has never been told before.

'Tremendous . . . moves me to bestow a reviewer's cliché I long ago vowed never to use: a *tour de force*' Robert Macfarlane, *Spectator*

'A magnificent history, enormously entertaining' *Daily Telegraph*

'A splendid book . . . exposes the factual roots of some of science's well known tales (for example, Galileo never dropped weights of different sizes from Pisa's leaning tower)' *Economist*

FREEDOM EVOLVES   DANIEL DENNETT

World-renowned philosopher Daniel Dennett shows how the traditional opposition between free will and determinism is misconceived: free will *does* exist. It is not an eternal, unchanging feature of our existence, such as the law of gravity – instead, freedom evolves. Like the planet's atmosphere on which life depends, the conditions on which our freedom depends had to evolve, and, like the atmosphere, they continue to evolve – and could be extinguished.

'Daniel Dennett's new book combines, once again, original philosophical thinking, marvellously vivid prose and extraordinarily lucid argumentation. *Freedom Evolves* does what I would have thought impossible: it says something new about free will and determinism' Richard Rorty

PENGUIN SCIENCE

ROBOT RODNEY A. BROOKS

'Brooks gives a wonderfully personal insight into the remarkable paradigm shift that AI is still undergoing' *Guardian*

We are constantly developing more and more sophisticated technology: the robot *Sojourner* explored the surface of Mars, and the prototype *Kismet* can learn to talk, recognise faces and socialise just like a child. With such complex technologies come questions: could we build a robot with thoughts and emotions? What would this mean for humanity? In this provocative and compelling book, Rodney Brooks, the world's leading robotics expert, not only explores the current state of robotics and artificial intelligence, but also conceives a radical alternative to established views of our cybernetic future.

'When Brooks tells me that in 20 more years we will have robots with feelings and consciousness, I'm not going to argue with him' *The New York Times*

'Excellent, highly readable. . . the best kind of popular science book: clear and accessible but not dumbed down' *American Scientist*

SIX EASY PIECES RICHARD P. FEYNMAN

Drawn from Richard Feynman's celebrated and landmark text 'Lecture on Physics', this collection of essays are the perfect example of his gift for making complex subjects accessible and hugely entertaining. They reveal Feynman's distinctive style while introducing the essentials of physics to the general reader.

'A delightful volume – it serves as both a primer on physics for non-scientists and as a primer on Feynman himself' Paul Davies

'He explores, in an accessible way, big problems about the difference between past and future, left and right, and the meaning of uncertainty. One of the most enjoyable books ever written by a major scientist'. *Observer*

'Fascinating. . . an insight into the thought processes of a great physicist' *The Times Literary Supplement*

PENGUIN SCIENCE

THE LANGUAGE INSTINCT   STEVEN PINKER

How did language evolve? How old is it? Why is structural grammatical language specifically human? In his illuminating book, MIT psychologist Steven Pinker attacks these fundamental questions with intense curiosity, energetic wit and clarity.

'A marvellously readable book . . . illuminates every facet of human language: its biological origin, its uniqueness to humanity, its acquisition by children, its grammatical structure, the production and perception of speech, the pathology of language disorders and its unstoppable evolution' *Nature*

'An extremely valuable book, informative and well written'   Noam Chomsky

HOW THE MIND WORKS   STEVEN PINKER

In his marvellously fun, awesomely informative survey of modern brain science, Pinker extends the Darwinian cognitive approach of his previous book to the mind in general, covering its aspects from vision, memory and consciousness to humour, fear, lust and anger.

'Why do memories fade? Why do we lose our tempers? Why do fools fall in love? Pinker's objective in this erudite account is to explore the nature and history of the human mind' Cheryl Younson, *Sunday Times*

'Witty popular science that you enjoy reading for the writing as well as for the science' *The New York Review of Books*

THE BLANK SLATE   STEVEN PINKER

With wit, lucidity and startling insight, Pinker explains how the new sciences of mind, brain, genes and evolution, far from being corrosive of humane values, complement observations about the human condition made by millennia of artists and philosophers.

'The best book on human nature that I or anyone else will ever read'   Matt Ridley

'A passionate defence of the enduring power of human nature. . .both life-affirming and deeply satisfying' Tim Lott, *Daily Telegraph*

'Brilliant. . .enjoyable, informative, clear, humane' *New Scientist*

PENGUIN SCIENCE

GENES, PEOPLES AND LANGUAGES
LUIGI LUCA CAVALLI-SFORZA

'It would be a slight exaggeration to say that Cavalli-Sforza studies everything about everybody, because actually he is "only" interested in what genes, languages, archaeology, and culture can teach us about the history and migrations of everybody for the last several hundred thousand years' Jared Diamond

THE JOURNEY OF MAN   SPENCER WELLS

'Packed with important insights into our history. . . Who needs literature when science is this much fun?' *Guardian*

Spencer Wells embarks on a unique voyage of discovery, travelling the world and deciphering the genetic codes of people from the Sahara Desert to Siberia. He reveals how our DNA enables us to work out where our ancestors lived and retraces their footsteps as they spread from Africa to the far corners of the earth.